PURE AND APPLIED MATHEMATICS

W9-ABR-768

THE AVERAGED MODULI
OF SMOOTHNESS

THE AVERAGED MODULI OF SMOOTHNESS

APPLICATIONS IN NUMERICAL METHODS AND APPROXIMATION

BLAGOVEST SENDOV
VASIL A. POPOV
Bulgarian Academy of Sciences

Translation edited by
G. M. Phillips
University of St Andrews

A Wiley–Interscience Publication

JOHN WILEY & SONS
Chichester · New York · Brisbane · Toronto · Singapore

First published as Usredneni Moduli na Gladkost,
Bulgarian Mathematical Monographs Volume 4,
by the Publishing House of the Bulgarian Academy
of Sciences, Sofia.

Library of Congress Cataloging-in-Publication Data:

Sendov, Blagovest.
    The averaged moduli of smoothness.

    (Pure and applied mathematics)
    Translation of: Usredneni moduli na gladkost.
    'A Wiley–Interscience publication.'
    Bibliography: p.
    1. Smoothness of functions.
I. Popov, Vasil A. (Vasil Atanasov)
II. Title.   III. Series: Pure and
applied mathematics (John Wiley & Sons)
QA355.S4613 1988   515'.223   88–2434
ISBN 0 471 91952 7

British Library Cataloguing in Publication Data:

Sendov, Blagovest
    The averaged moduli of smoothness.
    1. Numerical methods. Errors. Estimation
    I. Title II. Popov, Vasil A.   III. Series
    519.4

ISBN 0 471 91952 7

Phototypesetting by Thomson Press (India) Limited, New Delhi
Printed and bounded in Great Britain by Anchor Brendon Ltd, Tiptree, Essex

CONTENTS

PREFACE

The idea of writing this book arose from a new method of estimating the error in a large number of numerical processes such as interpolation, approximation of functions by means of operators, quadrature formulas, and network methods for the numerical solution of integral and differential equations. The method is based on the use of a new characteristic of functions, used for the first time in the theory of Hausdorff approximations, and subsequently in the classical theory of approximation of functions by means of operators.

We have given these new characteristics the name averaged moduli of smoothness, or τ-moduli. They are defined for every bounded function and are an integral analogue of the classical moduli of continuity and smoothness for the uniform (Chebyshev) metric, i.e. of the so-called ω-moduli.

In the theory of approximation the integral ω-moduli of continuity in L_p-spaces, $1 \leqslant p < \infty$, are used as another analogue of the ω-moduli in the uniform case.

The novelty in our approach lies in the method of obtaining an analogue of the uniform case.

It turns out that the τ-moduli, which have already been used with success in a series of problems in the theory of approximation, are particularly helpful for estimating the error of numerical methods which deal with the values of functions at a finite number of points (or potentially countably many).

The purpose of this book is to show, by a number of examples, the methods and results of applying the averaged moduli of smoothness in problems of error estimation in numerical methods and approximation.

The importance of the τ-moduli lies in the possibility of obtaining an estimate of the error in a given method of approximation of functions, or in a given numerical method, without additional restrictions on the functions, except those necessary for the formulation of the problem. Quadrature formulae are typical examples. For although they are defined for every Riemann integrable function, the usual error estimates require the existence and boundedness of the derivative of the function of some order.

By applying the τ-moduli, we can obtain estimates without additional assumptions. On the other hand, if we have additional information, we can obtain all known estimates, sometimes with weaker restrictions. Another important point is that the use of the averaged moduli does not only allow us to express the error in a different way; in many cases it leads to new results. The authors hope that the scope of this method is much wider than the examples given in the book.

The scheme of the book is as follows: in Chapter 1 the classical ω-moduli are described and the τ-moduli are introduced. Connections between the ω-moduli and τ-moduli are discussed.

In Chapter 2 the basic apparatus for applying the τ-moduli is given. This includes the theorem of H. Whitney, the interpolation theorem for the averaged moduli, the theorem for intermediate approximations and the Riesz–Thorin interpolation theorem.

In Chapter 3, estimates using τ-moduli for the most frequently used quadrature formulae are given. In Chapter 4 we treat the application of the averaged moduli of smoothness to the approximation of functions by means of summation operators (the Bernstein operators, general positive operators and spline-interpolation operators).

τ-moduli are used in Chapter 5 to obtain estimates for the numerical solution of the Fredholm equation of the second kind.

In Chapter 6, estimates using τ-moduli are obtained for the numerical solution of the initial value problem for ordinary first-order differential equations.

Chapter 7 describes the application of the averaged moduli to the numerical solution of the boundary value problem for linear second-order differential equations.

Finally, Chapter 8 presents the application of the averaged moduli of smoothness to one-sided trigonometrical approximation of functions. Direct and converse theorems are obtained in terms of the τ-moduli, as well as a characterization of the order of best one-sided trigonometrical approximation.

Acknowledgments. The authors wish to thank the following colleagues for their help: A. S. Andeev, P. P. Petrushev, K. G. Ivanov, S. P. Tashev, E. Moskona, P. Binev, Vl. Hristov, R. Maleev.

NOTATION

1. $C_{[a,b]} = C_\Delta$ the set of continuous functions on the interval $[a,b] = \Delta$
2. $M_{[a,b]} = M_\Delta$ the set of bounded and measurable functions on the interval $[a,b] = \Delta$
3. H_n the set of algebraic polynomials of degree at most n
4. $L_\infty[a,b] = L_\infty\Delta$ the set of all essentially bounded functions on $[a,b] = \Delta$
5. $L_p[a,b] = L_p\Delta$ the set of all measurable functions whose pth power is integrable in $[a,b] = \Delta$
6. $W_p^r[a,b] = W_p^r$ the set of all functions with an absolutely continuous $(r-1)$th derivative and with the rth derivative in $L_p[a,b]$
7. $\|f\|_{C[a,b]} = \|f\|_C = \sup\{|f(x)|: x\in[a,b], f\in M_{[a,b]}\}$
8. $\|f\|_{L_p[a,b]} = \|f\|_{L_p} = \{1/(-a)\int_a^b |f(x)|^p\, dx\}^{1/p}, p \geqslant 1$
9. $\|f\|_{L_\infty[a,b]} = \|f\|_{L_\infty} = \operatorname{ess\,sup}\{|f(x)|: x\in[a,b]\} = \lim_{p\to\infty} \|f\|_{L_p}$
10. $V_a^b f$ the variation of f on the interval $[a,b]$
11. $T_n(f)$ the algebraic polynomial of best uniform approximation to the function f of degree $\leqslant n$ on a given interval, $\|f - T_n(f)\|_C \leqslant \|f - p\|_C; p\in H_n$
12. $E_k(f) = \inf\{\|f - p\|_C: p\in H_k\}$ the best uniform approximation to the function f by algebraic polynomials of degree $\leqslant k$
13. $\omega(f;\delta) = \sup\{|f(x') - f(x'')|: |x' - x''| \leqslant \delta, x', x''\in[a,b]\}$ the modulus of continuity for the function f on the interval $[a,b]$
14. $\Delta_h^k f(x) = \sum_{m=0}^k (-1)^{m+k}\binom{k}{m} f(x + mh)$ the kth finite difference of the function f at the point x with step h
15. $\omega_k(f;\delta) = \sup\{|\Delta_h^k f(x)|: |h| \leqslant \delta, x, x + kh\in[a,b]\}$ the modulus of smoothness of order k for the function f on the interval $[a,b]$
16. $\omega_k(f;[a,b]) = \omega_k(f;\Delta) = \sup\{|\Delta_h^k f(x)|; x, x + kh\in[a,b] = \Delta\}$ the value of the modulus $\omega_k(f;\delta)$ of order k for the maximal value of $\delta = (b-a)/k$
17. $\omega_k(f;\delta)_{L_p} = \sup_{0\leqslant h\leqslant\delta} \{\int_a^{b-kh} |\Delta_h^k f(x)|^p dx\}^{1/p}$ the integral modulus of smoothness of order k
18. $\omega_k(f,x;\delta) = \sup\{|\Delta_h^k f(t)|: t, t + kh\in[x - k\delta/2, x + k\delta/2]\cap[a,b]\}$ the local modulus of smoothness of order k
19. $\tau_k(f;\delta)_{L_p} = \|\omega_k(f,\cdot;\delta)\|_{L_p}$ the averaged modulus of smoothness of order k
20. $\tau(f;\delta)_{L_p} = \tau_1(f;\delta)_{L_p}$
21. $\|f\|_{l_p\Sigma_n} = \{\sum_{i=1}^n |f(x_i)|^p\Delta_i\}^{1/p}$

$\Sigma_n = \{x_i : a \leqslant x_0 \leqslant x_1 \leqslant \cdots \leqslant x_{n+1} = b\}, \ \Delta_i = x_{i+1} - x_i$

the discrete l_p norm

22. $t_+ = \begin{cases} t, & t \geqslant 0 \\ 0, & t < 0 \end{cases}$

23. $Q^2[0, 1]$ the set of all continuous functions with a piecewise continuous second derivative

24. T_n the set of trigonometric polynomials of degree $\leqslant n$

1 MODULI OF FUNCTIONS

In mathematical analysis, and in particular in approximation theory, moduli of functions are used to characterize some properties of the function. In this chapter we shall consider the classical moduli of continuity and smoothness, their integral generalizations and the basis for our averaged moduli of smoothness.

1.1. Moduli of continuity and moduli of smoothness

Let the function f be defined in the interval $[a, b]$.

Definition 1.1. The modulus of continuity of the function f is the following function of $\delta \in [0, b - a]$:

$$\omega(f; \delta) = \sup\{|f(x) - f(x')|: |x - x'| \leqslant \delta, x, x' \in [a, b]\} \tag{1.1}$$

From the definition we see immediately that the modulus of continuity of the function f gives the maximum oscillation of f in an interval of length $\delta > 0$.

A necessary and sufficient condition for a function f to be continuous in the interval $[a, b]$ is

$$\lim_{\delta \to 0} \omega(f; \delta) = \omega(f; 0) = 0 \tag{1.2}$$

The moduli of smoothness form a natural generalization of the modulus of continuity.

For every function f we define the kth difference with step h at a point x as follows:

$$\Delta_h^k f(x) = \sum_{m=0}^{k} (-1)^{m+k} \binom{k}{m} f(x + mh) \quad \Delta_h f(x) = \Delta_h^1 f(x) \tag{1.3}$$

where

$$\binom{k}{m} = \frac{k!}{m!(k - m)!}$$

is the binomial coefficient.

Let the function f be defined and bounded in the interval $[a,b]$ and let k be a natural number.

Definition 1.2. The modulus of smoothness of order k of a function f is the following function of $\delta \in [0, (b-a)/k]$:

$$\omega_k(f; \delta) = \sup\{|\Delta_h^k f(x)|: |h| \leqslant \delta, x, x + kh \in [a,b]\} \tag{1.4}$$

We shall also use the notation

$$\omega_k(f; \Delta) = \omega_k(f; [a,b]) = \sup\{|\Delta_h^k f(x)|: x, x + kh \in [a,b] = \Delta\}$$

Evidently the modulus of smoothness of order 1 is precisely the modulus of continuity, i.e. $\omega_1(f; \delta) = \omega(f; \delta)$. Sometimes the modulus of smoothness of order 2, $\omega_2(f; \delta)$, is called simply the modulus of smoothness or Zigmund's modulus. The moduli of smoothness have the following five basic properties:

(1) Monotonicity:

$$\omega_k(f; \delta') \leqslant \omega_k(f; \delta'') \quad \text{for } 0 \leqslant \delta' \leqslant \delta''$$

(2) Semi-additivity:

$$\omega_k(f + g; \delta) \leqslant \omega_k(f; \delta) + \omega_k(g; \delta)$$

(3) A higher order modulus can be estimated by means of a modulus of lower order.

$$\omega_k(f; \delta) \leqslant 2\omega_{k-1}(f; \delta)$$

(4) A modulus of the function can be estimated by means of a lower order modulus of the derivative.

$$\omega_k(f; \delta) \leqslant \delta \omega_{k-1}(f'; \delta)$$

(5) We can have an integer multiplier before the modulus:

$$\omega_k(f; n\delta) \leqslant n^k \omega_k(f; \delta) \quad \text{where } n \text{ is a natural number}$$

From (1) and (5) we obtain the following property.

(5') We can have a multiplier before the modulus:

$$\omega_k(f; \lambda\delta) \leqslant (\lambda + 1)^k \omega_k(f; \delta) \quad \lambda > 0$$

The proofs of properties (1) and (2) follow immediately from Definition 1.2. Property (3) follows from the identity

$$\Delta_h^k f(x) = \Delta_h^{k-1} f(x + h) - \Delta_h^{k-1} f(x)$$

Property (4) holds for absolutely continuous functions with bounded first derivative. For every h, $|h| \leqslant \delta$, we have

$$|\Delta_h^k f(x)| = |\Delta_h^{k-1}[f(x + h) - f(x)]|$$

$$= \left| \Delta_h^{k-1} \int_0^h f'(x + t) \, dt \right|$$

$$= \left| \int_0^h \Delta_h^{k-1} f'(x+t) \, dt \right|$$

$$\leqslant \int_{\min(0,h)}^{\max(0,h)} |\Delta_h^{k-1} f'(x+t)| \, dt$$

$$\leqslant \int_{\min(0,h)}^{\max(0,h)} \omega_{k-1}(f';\delta) \, dt \leqslant \delta \omega_{k-1}(f';\delta)$$

therefore

$$\sup \{ |\Delta_h^k f(x)| : |h| \leqslant \delta, x, x + kh \in [a,b] \}$$
$$= \omega_k(f;\delta) \leqslant \delta \omega_{k-1}(f';\delta)$$

To prove property (5) we use the identity

$$\Delta_{nh}^k f(x) = \sum_{i_1=0}^{n-1} \sum_{i_2=0}^{n-1} \cdots \sum_{i_k=0}^{n-1} \Delta_h^k f(x + i_1 h + i_2 h + \cdots + i_k h) \qquad (1.5)$$

We can prove identity (1.5) by induction on k. For $k = 1$ we have

$$\Delta_{nh} f(x) = f(x + nh) - f(x)$$

$$= \sum_{i=0}^{n-1} [f(x + ih + h) - f(x + ih)]$$

$$= \sum_{i=0}^{n-1} \Delta_h f(x + ih)$$

Let us suppose that (1.5) is valid for a given natural number k. Then

$$\Delta_{nh}^{k+1} f(x) = \Delta_{nh}^k [f(x + nh) - f(x)]$$

$$= \Delta_{nh}^k [\Delta_{nh} f(x)]$$

$$= \sum_{i_1=0}^{n-1} \sum_{i_2=0}^{n-1} \cdots \sum_{i_k=0}^{n-1} \Delta_h^k [\Delta_{nh} f(x + i_1 h + i_2 h + \cdots + i_k h)]$$

$$= \sum_{i_1=0}^{n-1} \sum_{i_2=0}^{n-1} \cdots \sum_{i_k=0}^{n-1} \Delta_h^k \left[\sum_{i_{k+1}=0}^{n-1} \Delta_h f(x + i_1 h + \cdots + i_k h + i_{k+1} h) \right]$$

$$= \sum_{i_1=0}^{n-1} \sum_{i_2=0}^{n-1} \cdots \sum_{i_{k+1}=0}^{n-1} \Delta_h^{k+1} f(x + i_1 h + i_2 h + \cdots + i_{k+1} h)$$

From (1.5) we obtain for h, $|h| \leqslant \delta$,

$$|\Delta_{nh}^k f(x)| \leqslant \sum_{i_2=0}^{n-1} \cdots \sum_{i_k=0}^{n-1} |\Delta_h^k f(x + i_1 h + \cdots + i_k h)|$$

$$\leqslant n^k \omega_k(f;\delta)$$

which is sufficient to prove property (5).

The modulus of continuity $\omega(f;\delta)$ has the following additional property:

(6) We can estimate $\omega(f;\delta)$ by means of the norm of the derivative:

$$\omega(f;\delta) \leqslant \delta \|f'\|_{C[a,b]}$$

Proof. We have for every function f, $f' \in C_{[a,b]}$

$$|f(x+h)-f(x)| = \left|\int_x^{x+h} f'(t)\,dt\right|$$

$$\leqslant |h|\|f'\|_{C[a,b]} \tag{1.6}$$

Applying this inequality we obtain property (6). ∎

More generally, if a function f has a bounded kth derivative, from properties (4) and (6) we obtain the following property of $\omega_k(f;\delta)$:

(6′) The kth modulus of smoothness can be estimated by means of the norm of the kth derivative:

$$\omega_k(f;\delta) \leqslant \delta^k \|f^{(k)}\|_{C[a,b]} \tag{1.7}$$

Property (3) shows that $\omega_k(f;\delta)$ can be estimated by means of $\omega_i(f;\delta)$ for every $i < k$. The converse is more complicated and is given by the following statement of **Marchaud**, which we shall give without proof.

Theorem of Marchaud. *For every $i < k$ we have the inequality*

$$\omega_i(f;\delta) \leqslant c_k \delta^i \left[\int_\delta^{(b-a)/k} t^{-i-1}\omega_k(f;t)\,dt + (b-a)^{-i}\|f\|_{C[a,b]} \right]$$

where c_k is a constant depending only on k (see Dzjadyk [1] or Timan [1]).

1.2.   Integral moduli

Using integral norms, we can obtain analogues of the moduli of smoothness, which are called integral moduli of smoothness, or simply integral moduli.

Let $f \in L_p[a,b]$, $1 \leqslant p \leqslant \infty$.

Definition 1.3. The integral modulus (L_p-modulus or p-modulus) of order k of the function f is the following function of $\delta \in [0, (b-a)/k]$:

$$\omega_k(f;\delta)_{L_p} = \sup_{0 \leqslant h \leqslant \delta} \left\{ \int_a^{b-kh} |\Delta_h^k f(x)|^p dx \right\}^{1/p} \tag{1.8}$$

The integral moduli have properties which are analogous to properties (1)–(6) of the moduli of smoothness.

We shall prove here one specific property of $\omega_1(f;\delta)_{L_1} = \omega(f;\delta)_L$ in the case where f is of bounded variation:

(7) We can estimate $\omega(f;\delta)_L$ by means of the variation of f:

$$\omega(f;\delta)_L \leqslant \delta V_a^b f \tag{1.9}$$

Proof. We have

$$\int_a^{b-h} |f(x+h) - f(x)| \, dx \leqslant \int_a^{b-h} V_x^{x+h} f \, dx$$

$$= \int_a^{b-h} (V_a^{x+h} f - V_a^x f) \, dx$$

$$= \int_{a+h}^b V_a^x f \, dx - \int_a^{b-h} V_a^x f \, dx$$

$$\leqslant \int_{b-h}^b V_a^x f \, dx$$

$$\leqslant h V_a^b f$$

From these inequalities we obtain (1.9). ∎

It is important to know we can invert the inequality (1.9). More precisely, the following theorem of Hardy and Littlewood holds:

Theorem 1.1. (*Hardy and Littlewood*). *Let $\omega(f; \delta)_L = O(\delta)$. Then the function f is equivalent to a function with bounded variation.*

Proof. Let $f \in L[a, b]$ and $\omega(f; \delta)_L \leqslant M\delta$. Let $c = (a + b)/2$ and

$$f_h(x) = \begin{cases} h^{-1} \int_0^h f(x + t) \, dt & \text{for } x \in [a, c] \\ -h^{-1} \int_0^{-h} f(x + t) \, dt & \text{for } x \in (c, b] \end{cases}$$

where $h \in (0, (b - a)/2)$. Since $f \in L[a, b]$, for every $x \in [a, b] \backslash Q$, where $Q \subset [a, b]$ is a set with measure zero, we have

$$\lim_{h \to +0} f_h(x) = f(x) \tag{1.10}$$

For the variation of f_h we have

$$V_a^c f_h = \int_a^c |f_h'(x)| \, dx$$

$$= h^{-1} \int_a^c |f(x + h) - f(x)| \, dx$$

$$\leqslant h^{-1}(b - a)\omega(f; h)_L \leqslant M$$

Analogously,

$$V_{c+0}^b f_h \leqslant M$$

On the other hand

$$|f_h(c) - f_h(c+0)| = \left| h^{-1} \int_0^h f(c+t)\,dt - \lim_{x \to c+0} h^{-1} \int_{-h}^0 f(x+t)\,dt \right|$$

$$\leqslant h^{-1} \int_0^h |f(c+t) - f(c+t-h)|\,dt$$

$$\leqslant h^{-1} \omega(f;h)_L \leqslant M$$

Since

$$V_a^b f_h = V_a^c f_h + |f_h(c) - f_h(c+0)| + V_{c+0}^b f_h$$

from the estimations given above we obtain

$$V_a^b f_h \leqslant 3M \tag{1.11}$$

Let x_i, $i = 1, \ldots, N$, $x_1 < x_2 < \cdots < x_N$, be arbitrary points in $[a,b] \backslash Q$. From (1.10) and (1.11) we obtain

$$\sum_{i=1}^{N-1} |f(x_{i+1}) - f(x_i)| = \lim_{h \to 0} \sum_{i=1}^{N-1} |f_h(x_{i+1}) - f_h(x_i)|$$

$$\leqslant \lim_{h \to 0} V_a^b f_h \leqslant 3M$$

Therefore f is of bounded variation on the set $[a,b] \backslash Q$, and, for $x \in [a,b] \backslash Q$, the function f can be represented in the form $f = f_1 - f_2$, where f_1 and f_2 are monotone increasing (non-decreasing) functions on $[a,b] \backslash Q$. Since Q has measure zero, $[a,b] \backslash Q$ is dense in $[a,b]$. Therefore we can define functions f_1 and f_2 on Q by continuity on the left, for example. Thus we obtain a function $\tilde{f} = f_1 - f_2$ defined in the whole interval $[a,b]$, which is of bounded variation on $[a,b]$, and which coincides with f almost everywhere. ∎

1.3.   Averaged moduli

The moduli of smoothness, which we have defined in Section 1.1, can also be defined using the local moduli of smoothness.

Let the function f be defined and bounded in the interval $[a,b]$.

Definition 1.4.   The local modulus of smoothness of the function f of order k at a point $x \in [a,b]$ is the following function of $\delta \in [0, (b-a)/k]$:

$$\omega_k(f, x; \delta) = \sup \left\{ |\Delta_h^k f(t)| : t, t + kh \in \left[x - \frac{k\delta}{2}, x + \frac{k\delta}{2} \right] \cap [a,b] \right\}$$

The function $\omega_k(f, x; \delta)$ as a function of x is defined for every $x \in [a,b]$, and obviously

$$\omega_k(f; \delta) = \| \omega_k(f, \cdot; \delta) \|_{C[a,b]}$$

This new definition of the moduli of smoothness leads us to some further moduli, which can be obtained from the local moduli using integral norms instead of the uniform norm.

Definition 1.5. The averaged modulus of smoothness of order k (or τ-modulus) of the function $f \in M[a,b]$ (see Notation 2) is the following function of $\delta \in [0, (b-a)/k]$:

$$\tau_k(f;\delta)_p = \| \omega_k(f,\cdot;\delta) \|_{L_p}$$

$$= \left[\int_a^b (\omega_k(f,x;\delta))^p \, dx \right]^{1/p}$$

This definition is correct, and we shall prove later that for every function $f \in M[a,b]$ the function $\omega_k(f,x;\delta)$ as a function of x also belongs to $M[a,b]$ and therefore belongs to $L_p[a,b]$, $1 \leq p \leq \infty$, (see Theorem 1.3).

For $k = 1$, instead of $\tau_1(f;\delta)_p$ we often write $\tau(f;\delta)_p$.

The averaged modulus $\tau(f;\delta)_p$ can also be defined in the following (equivalent) way. Let us denote

$$S(f,\delta;x) = \sup \left\{ f(t) : t \in \left[x - \frac{\delta}{2}, x + \frac{\delta}{2} \right] \cap [a,b] \right\}$$

$$I(f,\delta;x) = \inf \left\{ f(t) : t \in \left[x - \frac{\delta}{2}, x + \frac{\delta}{2} \right] \cap [a,b] \right\}$$

Then the following equality holds:

$$\omega_1(f,x;\delta) = S(f,\delta;x) - I(f,\delta;x)$$

and therefore

$$\tau(f;\delta)_p = \| S(f,\delta;\cdot) - I(f,\delta;\cdot) \|_{L_p[a,b]} \tag{1.12}$$

i.e. $\tau(f;\delta)_p$ is equal to the L_p-distance between the functions $S(f,\delta)$ and $I(f,\delta)$.

This definition of $\tau(f;\delta)_p$ allows us to introduce a modulus of continuity in every metric space of functions. If the metric is r, then we can set

$$\tau(f;\delta)_r = r(S(f,\delta), I(f,\delta)) \tag{1.12'}$$

The τ-moduli was first introduced using equation (1.12') (see Sendov [1] and Korovkin [2]). These moduli were used to investigate the convergence of sequences of linear positive operators with respect to different metrics (Sendov [2]). Dolzenko and Sevastjanov [1] independently introduced $\tau(f;\delta)_L$ and used this modulus to estimate the best piecewise monotone approximations with respect to the Hausdorff distance (see Sendov [3]).

The averaged moduli have five basic properties, which are analogous to be basic properties of the moduli of smoothness:

(1) Monotonicity:

$$\tau_k(f;\delta')_p \leq \tau_k(f;\delta'')_p \quad \text{for } \delta' \leq \delta''$$

(2) Semi-additivity:

$$\tau_k(f+g;\delta)_p \leqslant \tau_k(f;\delta)_p + \tau_k(g;\delta)_p$$

(3) A higher-order-modulus can be estimated by means of a modulus of lower order:

$$\tau_k(f;\delta)_p \leqslant 2\tau_{k-1}\left(f;\frac{k}{k-1}\delta\right)_p$$

(4) The modulus of order k of the function can be estimated from the modulus of order $k-1$ of the derivative:

$$\tau_k(f;\delta)_p \leqslant \delta\tau_{k-1}\left(f';\frac{k}{k-1}\delta\right)_p$$

(5) We can take an integer multiplier before the modulus:

$$\tau_k(f;n\delta)_p \leqslant (2n)^{k+1}\tau_k(f;\delta)_p$$

where n is a natural number.

Properties (1)–(3) of the averaged moduli of smoothness are exactly the same as the corresponding properties of the usual moduli of smoothness and the proofs of these properties are exactly like those for the moduli of smoothness. Properties (4) and (5) are a little different.

To prove (4) we shall use the identity

$$\Delta_h^k f(t) = \int_0^h \Delta_h^{k-1} f'(t+u)\,du \quad h > 0 \tag{1.13}$$

From (1.13) we obtain

$$\sup\left\{|\Delta_h^k f(t)|:t, t+kh\in\left[x-\frac{k\delta}{2}, x+\frac{k\delta}{2}\right]\cap[a,b]\right\}$$

$$\leqslant \sup\left\{\int_0^h |\Delta_h^{k-1} f'(t+u)|\,du: t, t+kh\in\left[x-\frac{k\delta}{2}, x+\frac{k\delta}{2}\right]\cap[a,b]\right\} \tag{1.14}$$

If

$$t, t+kh\in\left[x-\frac{k\delta}{2}, x+\frac{k\delta}{2}\right]\cap[a,b]$$

and $h > 0$, then the points $t+u$, $t+u+(k-1)h$ belong to the same interval for $0 \leqslant u \leqslant h$. Therefore we have, from (1.14),

$$|\Delta_h^{k-1} f'(t+u)| \leqslant \omega_{k-1}(f',x;\delta')$$

where

$$\delta' = \frac{k}{k-1}\delta$$

thus

$$\omega_k(f,x;\delta) \leqslant \delta\omega_{k-1}\left(f',x;\frac{k}{k-1}\delta\right) \quad x\in[a,b]$$

If we take the L_p norm of both sides of this inequality we obtain property (4). To prove property (5) we use the identity

$$\Delta_{nh}^k f(t) = \sum_{i=0}^{(n-1)k} A_i^{n,k} \Delta_h^k f(t+ih) \tag{1.15}$$

where $A_i^{n,k}$ are defined by

$$(1+t+\cdots+t^{n-1})^k = \sum_{i=0}^{(n-1)k} A_i^{n,k} t^i$$

Identity (1.15) follows from (1.5). Obviously $A_i^{n,k} > 0$ for $i = 0, 1, \ldots, (n-1)k$ and we have the equation

$$\sum_{i=0}^{(n-1)k} A_i^{n,k} = n^k \tag{1.16}$$

From (1.15) we obtain

$$|\Delta_{2nh}^k f(t)| \leqslant \sum_{i=0}^{(2n-1)k} A_i^{2n,k} |\Delta_h^k f(t+ih)| \tag{1.17}$$

If

$$t, t + 2nkh \in \left[x - \frac{kn\delta}{2}, x + \frac{kn\delta}{2} \right] \cap [a,b]$$

then the points $t + ih$, $t + ih + kh$ belong to one of the intervals

$$\left[x - \frac{kn\delta}{2} + (j-1)\frac{k\delta}{2}, \quad x - \frac{kn\delta}{2} + (j+1)\frac{k\delta}{2} \right] j = 1, 2, \ldots, 2n-1$$

Therefore we obtain

$$|\Delta_{2nh}^2 f(t)| \leqslant \sum_{i=0}^{(2n-1)k} A_i^{2n,k} \sum_{j=1}^{2n-1}{}' \omega_k\left(f, x - (n-j)\frac{k\delta}{2}; \delta \right)$$

or

$$\omega_k(f, x; n\delta) \leqslant \sum_{i=0}^{(2n-1)k} A_i^{2n,k} \sum_{j=1}^{2n-1}{}' \omega_k\left(f, x - (n-j)\frac{k\delta}{2}; \delta \right) \tag{1.18}$$

where the only terms to appear in the sum $\sum'$ are those for which $x - (n-j)k\delta/2 \in [a,b]$. If we take the L_p-norm of both sides of (1.18), we obtain using equation (1.16)

$$\tau_k(f; n\delta)_p \leqslant (2n)^k (2n-1) \tau_k(f; \delta)_p$$

Property 5 follows.

From properties (5) and (1) we obtain the following property.

(5′) We can take a multiplier before the modulus:

$$\tau_k(f; \lambda\delta)_p \leqslant (2(\lambda+1))^{k+1} \tau_k(f; \delta)_p \quad \lambda > 0$$

In the case $k = 1$ the τ-moduli have some specific properties. The first of these is exactly the same as property (6) of the modulus of continuity.

(6) We can estimate the modulus using the norm of the derivative

$$\tau(f;\delta)_p \leqslant \delta \| f' \|_{L_p}$$

Proof. Let us extend f outside the interval $[a,b]$, setting $f(x)=f(a)$ for $x < a$ and $f(x)=f(b)$ for $x > b$. Then for every $x \in [a,b]$ we have

$$\omega(f,x;\delta) = \sup \left\{ |f(t') - f(t'')| : t', t'' \in \left[x - \frac{\delta}{2}, x + \frac{\delta}{2} \right] \right\}$$

$$= \sup \left\{ \left| \int_{t'}^{t''} f'(t)\,dt \right| : t', t'' \in \left[x - \frac{\delta}{2}, x + \frac{\delta}{2} \right] \right\}$$

$$\leqslant \int_{x-\delta/2}^{x+\delta/2} |f'(t)|\,dt$$

$$= \int_{-\delta/2}^{\delta/2} |f'(x+t)|\,dt$$

From this inequality, taking the L_p-norm, we obtain

$$\tau(f;\delta)_p = \| \omega(f,\cdot;\delta) \|_{L_p}$$

$$\leqslant \int_{-\delta/2}^{\delta/2} \| f'(\cdot + t) \|_{L_p}\,dt$$

$$= \delta \| f' \|_{L_p} \qquad \blacksquare$$

More generally, if the function f has a bounded derivative of order k, from properties (4) and (6) we obtain the following property of $\tau_k(f;\delta)_p$.

(6′) The kth modulus can be estimated by means of the norm of the kth derivative of the function

$$\tau_k(f;\delta)_p \leqslant c(k)\delta^k \| f^{(k)} \|_{L_p}$$

The second specific property is analogous to property (7) of the integral modulus $\omega(f;\delta)_L$

(7) We can estimate $\tau(f;\delta)_L$ using the variation of the function

$$\tau(f;\delta)_L \leqslant \delta V_a^b f$$

Proof. Again, let $f(x)=f(a)$, $x < a$, $f(x)=f(b)$, $x > b$. Then

$$\omega(f,x;\delta) \leqslant V_{x-\delta/2}^{x+\delta/2} f$$

and therefore

$$\tau(f;\delta)_L \leqslant \int_a^b V_{x-\delta/2}^{x+\delta/2} f\,dx$$

$$= \int_a^b V_{a-\delta/2}^{x+\delta/2} f\,dx - \int_a^b V_{a-\delta/2}^{x-\delta/2} f\,dx$$

$$= \int_{a+\delta/2}^{b+\delta/2} V_{a-\delta/2}^t f \, dt - \int_{a-\delta/2}^{b-\delta/2} V_{a-\delta/2}^t f \, dt$$

$$= \int_{b-\delta/2}^{b+\delta/2} V_{a-\delta/2}^t f \, dt - \int_{a-\delta/2}^{a+\delta/2} V_{\alpha-\delta/2}^t f \, dt$$

$$\leqslant \int_{b-\delta/2}^{b+\delta/2} V_a^b f \, dt = \delta V_a^b f \qquad\qquad \blacksquare$$

Property (7) is 'geometrically obvious from equation (1.12) for $\tau(f;\delta)_L$, since $\tau(f;\delta)_L$ is the area of the strip between $S(f;\delta)$ and $I(f;\delta)$ and this area cannot be bigger than the variation of f (the 'height' of the strip), multiplied by δ (the 'width' of the strip).

Now we shall prove a characteristic property of the modulus $\tau(f;\delta)_L$.

Theorem 1.2. *A necessary and sufficient condition for the function f to be Riemann integrable is*

$$\lim_{\delta \to +0} \tau(f;\delta)_L = 0 \qquad\qquad (1.19)$$

Proof. Using equation (1.12) for $\tau(f;\delta)_L$ we have

$$\tau(f;\delta)_L = \int_a^b [S(f,\delta;x) - I(f,\delta;x)] \, dx$$

It is easy to see that

$$\lim_{\delta \to 0} \int_a^b S(f,\delta;x) \, dx = \int_a^{\overline{b}} f(x) \, dx$$

where $\int_a^{\overline{b}}$ is the upper Darboux integral for the function f, and

$$\lim_{\delta \to 0} \int_a^b I(f,\delta;x) \, dx = \int_{\underline{a}}^b f(x) \, dx$$

is the lower Darboux integral for f. Therefore equation (1.19) is equivalent to

$$\int_a^{\overline{b}} f(x) \, dx = \int_{\underline{a}}^b f(x) \, dx$$

But this equation is a necessary and sufficient condition for function f to be Riemann integrable. $\blacksquare$

The estimate for the average moduli given by property (5) is not exact. We can obtain an exact estimate where $k = 1$.

(5″) We can have an integer multiplier before the modulus:

$$\tau(f;n\delta)_p \leqslant n\tau(f;\delta)_p$$

Proof. We have

$$\omega_1(f, x; n\delta) = \sup\left\{|f(t) - f(t+h)| : t, t+h \in \left[x - \frac{n\delta}{2}, x + \frac{n\delta}{2}\right] \cap [a, b]\right\}$$

Let us define

$$\xi_i(x) = x - (n - 2i + 1)\frac{\delta}{2} \quad i = 1, \ldots, n$$

We have

$$\omega_1(f, x; n\delta) \leqslant \sum_{i=1}^{n} w_1(f, \xi_i(x); \delta)$$

where the summation is taken over those i for which $\xi_i(x) \in [a, b]$. Therefore

$$\tau(f; n\delta)_p \leqslant \sum_{i=1}^{n} \|\omega_1(f, \xi_i(x); \delta)\|_p$$

$$\leqslant n\tau(f; \delta)_p$$

since

$$\|\omega_1(f, \xi_i(x); \delta)\|_p = \left\{\int_a^b \left(\omega_1\left(f, x - \frac{n - 2i + 1}{2}\delta; \delta\right)\right)^p dx\right\}^{1/p}$$

$$\leqslant \tau(f; \delta)_p \qquad\qquad\blacksquare$$

For the averaged moduli of smoothness we also have the Marchaud inequality:

$$\tau_k(f; \delta)_p \leqslant C(k)\delta^k\left[\int_\delta^{(b-a)/(k+1)} \frac{\tau_{k+1}(f; t)_p}{t^{k+1}} dt + \frac{\|f\|_{C[a,b]}}{(b-a)^k}\right]$$

To end this section, we shall give Ivanov's proof that for $f \in M[a, b]$ we have $\omega_k(f, x; \delta) \in M[a, b]$ as a function of x. We shall need some definitions and lemmas.

Definition 1.6.   We say that the set $X \subset [a, b]$ is closed on the left (on the right) if the limit of every monotone increasing (decreasing) sequence of points belonging to X also belongs to X.

Lemma 1.1.   *If $X \subset [a, b]$ is closed on the left (on the right), then X is measurable.*

Proof.   Let the set $X \subset [a, b]$ be closed on the left and $y \in Y = [a, b] \backslash X$. We shall prove that there exists $\varepsilon > 0$ such that $[y - \varepsilon, y] \subset Y$. Let us suppose that for every $\varepsilon > 0$ the interval $[y - \varepsilon, y]$ is not a subset of Y i.e. there exists $x_\varepsilon \in [y - \varepsilon, y]$ such that $x_\varepsilon \in X$. Therefore there exists a sequence $x_n \in X$, $x_1 \leqslant x_2 \leqslant \cdots \leqslant x_n \leqslant \cdots$, $\lim_{n \to \infty} x_n = Y$. Since X is closed on the left, we must have $y \in X$. which contradicts the assumption that $y \in Y$. Therefore the components of connectedness of Y are intervals, which are not degenerate at points, i.e. Y is a countable union of intervals and therefore X is measurable. $\blacksquare$

Lemma 1.2. *Let g be a function from $(-\infty, \infty) \times [0, \infty)$ to $(-\infty, \infty)$, $\delta > 0$, and*

$$A_x = A_x(\delta) = \left\{(t,h): t\in[x-\delta, x+\delta), h\in\left[0, \frac{x+\delta-t}{k}\right)\right\}$$

$$B_x = B_x(\delta) = \left\{(t,h): t\in(x-\delta, x+\delta], h\in\left(0, \frac{x+\delta-t}{k}\right]\right\}$$

where k is a fixed natural number.
    Then the function of x, $x\in[a,b]$,

$$G_A(x) = \sup\{g(t,h): (t,h)\in A_x\}$$
$$G_B(x) = \sup\{g(t,h): (t,h)\in B_x\}$$

are measurable functions.

Proof. Let c be an arbitrary real number and $F_c = \{x: x\in[a,b], G_A(x) \le c\}$. We shall prove that F_c is closed on the left, from which, using Lemma 1.1, it follows that F_c is measurable, i.e. the function G_A is measurable.
    Let $x_n\in F_c$, $x_n\le x_{n+1}\le x_0$, $\lim_{n\to\infty} x_n = x_0$. If $\varepsilon > 0$, then there exists an ordered pair $(t,h)\in A_{x_0}$ such that $g(t,h) > G_A(x_0) - \varepsilon$. Let N be sufficiently large that $x_0 - \theta < x_N \le x_0$, where $\theta = x_0 + \delta - (t+kh) > 0$. Then $(t,h)\in A_{x_N}$ since

$$x_N - \delta \le x_0 - \delta \le t = x_0 - \theta + \delta - kh \le x_0 - \theta + \delta < x_N + \delta$$
$$0 \le h = \frac{x_0 - \theta + \delta - t}{k} < \frac{x_N + \delta - t}{k}$$

Therefore

$$G_A(x_0) - \varepsilon < g(t,h) \le G_A(x_N) \le c$$

i.e. $G_A(x_0) \le c+\varepsilon$. Since $\varepsilon > 0$ is arbitrary, we have $G_A(x_0) \le c$, i.e. $x_0\in F_c$.
    In the same way we can prove that the function G_B is measurable. ∎

Theorem 1.3. *Let k be any natural number, $\delta > 0$ and $f\in M[a,b]$. Then $\omega_k(f,x;\delta)$ as a function of $x\in[a,b]$ belongs to $M[a,b]$.*

Proof. Let $\Delta_h^k f(t) = 0$ if t or $t+kh\notin[a,b]$. Let us define $g(t,h) = |\Delta_h^k f(t)|$. Using the notation of Lemma 1.2, let

$$A_x = A_x\left(\frac{k\delta}{2}\right), \quad B_x = B_x\left(\frac{k\delta}{2}\right)$$

Then

$$\omega_k(f,x;\delta) = \sup\left\{|\Delta_h^k f(t)|: t, t+kh\in\left[x-\frac{k\delta}{2}, x+\frac{k\delta}{2}\right]\cap[a,b]\right\}$$

$$= \sup\left\{g(t,h): t, t+kh\in\left[x-\frac{k\delta}{2}, x+\frac{k\delta}{2}\right], h\ge 0\right\}$$

$$= \max\left\{g\left(x-\frac{k\delta}{2}, \delta\right), G_A(x), G_B(x)\right\}$$

Functions G_A and G_B are measurable by Lemma 1.2, and the function

$$g\left(x - \frac{k\delta}{2}, \delta\right)$$

$$= \begin{cases} \left| \sum_{m=0}^{k} (-1)^{k+m} \binom{k}{m} f\left(x - \frac{k\delta}{2} + m\delta\right) \right| & \text{for } x \in \left[a + \frac{k\delta}{2}, b - \frac{k\delta}{2}\right] \\ 0 & \text{for } x \in [a,b] \setminus \left[a + \frac{k\delta}{2}, b - \frac{k\delta}{2}\right] \end{cases}$$

is measurable, since f is a measurable function.

Therefore $\omega_k(f, x; \delta)$, as a maximum of three measurable functions, is also measurable. ∎

Corollary 1.1. For every natural number k and $\delta > 0$, if $f \in M[a,b]$, then $\omega_k(f, x; \delta) \in L_p[a,b]$ as a function of x, $1 \leqslant p \leqslant \infty$.

1.4. Connections between the different types of moduli

Between the moduli of smoothness, the integral moduli and the averaged moduli there exists some order. The moduli of smoothness are the strongest and the integral moduli are the weakest. More precisely, the following statement holds:

Theorem 1.4. If f is a measurable bounded function on $[a,b]$, then

$$\omega_k(f; \delta)_p \leqslant \tau_k(f; \delta)_p \leqslant \omega_k(f; \delta)(b - a)^{1/p} \tag{1.20}$$

Proof. The second inequality follows immediately from the inequality

$$\|\omega_k(f, \cdot; \delta)\|_{L_p[a,b]} \leqslant \|\omega_k(f, \cdot; \delta)\|_{C[a,b]} (b - a)^{1/p}$$

We can obtain the first inequality from

$$\omega_k(f; \delta)_p = \sup_{0 \leqslant h \leqslant \delta} \left\{ \int_a^{b-kh} |\Delta_h^k f(x)|^p \, dx \right\}^{1/p}$$

$$\leqslant \sup_{0 < h \leqslant \delta} \left\{ \int_a^{b-kh} \left(\omega_k\left(f, x + \frac{kh}{2}; \delta\right) \right)^p dx \right\}^{1/p}$$

$$= \sup_{0 < h \leqslant \delta} \left\{ \int_{a+kh/2}^{b-kh/2} (\omega_k(f, x; \delta))^p \, dx \right\}^{1/p}$$

$$\leqslant \sup_{0 < h \leqslant \delta} \tau_k(f; \delta)_p$$

$$= \tau_k(f; \delta)_p$$

The inequality (1.20) shows that (1.9) is a consequence of property (7). ∎

The following important and non-trivial connection between the integral moduli and the averaged moduli exists for sufficiently smooth functions:

Theorem 1.5. *There exists a constant $c(k)$, depending only on $k \geqslant 2$, such that for every absolutely continuous function f on the interval $[a, b]$ we have*

$$\tau_k(f; \delta)_p \leqslant c(k) \delta \omega_{k-1}(f'; \delta)_p \qquad (1.21)$$

To prove this theorem we shall need two lemmas.

Lemma 1.3. *Let a function g be defined for all real x. Then for every natural number k and every choice of the real numbers h and v we have the identity:*

$$\Delta_h^k g(x) = \sum_{i=1}^{k} (-1)^i \binom{k}{i} [\Delta_{i(v-h)/k}^k g(x + ih) - \Delta_{h+i(v-h)/k}^k g(x)] \qquad (1.22)$$

Proof. We have

$$\sum_{l=0}^{k} (-1)^{k-l} \binom{k}{l} \Delta_{h+l(v-h)/k}^k g(x)$$

$$= \sum_{l=0}^{k} (-1)^{k-l} \binom{k}{l} \sum_{i=0}^{k} (-1)^{k-i} \binom{k}{i} g\left\{x + i\left[h + \frac{l(v-h)}{k}\right]\right\}$$

$$= \sum_{i=0}^{k} (-1)^{k-i} \binom{k}{i} \sum_{l=0}^{k} (-1)^{k-l} \binom{k}{l} g\left[x + ih + \frac{il(v-h)}{k}\right]$$

$$= \sum_{i=1}^{k} (-1)^{k-i} \binom{k}{i} \Delta_{i(v-h)/k}^k g(x + ih)$$

If we retain on the left-hand side of this equation only the term corresponding to $l = 0$, taking all other terms to the right-hand side, we obtain identity (1.22).

∎

Lemma 1.4. *If $g \in L[a, b]$, $t \in (a, b)$, k is a natural number, $\delta > 0$, $s > 0$ and $[t - k\delta, t + 2k\delta + s] \subset [a, b]$, then the following inequality holds:*

$$\sup_{0 < h \leqslant \delta} \int_0^s |\Delta_h^k g(t + u)| \, du \leqslant \frac{(k+1)2^{k+1}}{\delta} \int_{-\delta}^{\delta} \int_0^{s+k\delta} |\Delta_v^k g(t + u)| \, du \, dv$$

Proof. From the conditions of the lemma it follows that $\delta \leqslant (b - a)/3k$. From (1.22) we obtain, for every $x \in [a, b - kh]$, $h \in [0, \delta]$:

$$|\Delta_h^k g(x)| \leqslant \sum_{i=1}^{k} \binom{k}{i} \{|\Delta_{i(v-h)/k}^k g(x + ih)| + |\Delta_{h+i(v-h)/k}^k g(x)|\} \qquad (1.23)$$

We shall consider two cases:
(a) If $x \in [a, (a + b)/2]$, then $x + kh \leqslant x + k\delta \leqslant b$, and for every $v \in [0, \delta]$ we have

the inequalities

$$x + ih + \frac{ik(v - h)}{k} = x + iv \leqslant b$$

$$x + iv \geqslant a$$

and

$$x + k\left[h + \frac{i(v - h)}{k}\right] = x + (k - i)h + iv \leqslant x + k\delta \leqslant b$$

Therefore all finite differences on the right-hand side of (1.23) are defined. In this case we integrate both sides of (1.23) with respect to v on the interval $[0, \delta]$. Dividing both sides by δ, we obtain

$$|\Delta_h^k g(x)| \leqslant \frac{1}{\delta} \sum_{i=1}^{k} \binom{k}{i} \left\{ \int_0^\delta [|\Delta_{i(v-h)/k}^k g(x + ih) + |\Delta_{h+i(v-h)/k}^k g(x)|] \, dv \right\} \qquad (1.24)$$

(b) If $x + kh \in [(a + b)/2, b]$, we use the obvious equality

$$|\Delta_h^k g(x)| = |\Delta_{-h}^k g(x + kh)|$$

and from (1.23) we obtain

$$|\Delta_h^k g(x)| \leqslant \sum_{i=1}^{k} \binom{k}{i} \{|\Delta_{i(v+h)/k}^k g[x + (k - i)h]| + |\Delta_{h+i(v+h)/k}^k g(x + kh)|\} \qquad (1.25)$$

In this case we take $v \in [-\delta, 0]$. Then the following inequalities hold:

$$x + (k - i)h + \frac{ki(v + h)}{k} = x + k(h + v) \leqslant x + kh \leqslant b$$

$$x + k(h + v) \geqslant x - k\delta \geqslant a$$

$$x + kh + k\left[-h + \frac{i(v + h)}{k}\right] = x + i(v + h) \leqslant x + kh \leqslant b$$

$$x + i(v + h) \geqslant x - k\delta \geqslant a$$

Therefore in this case all finite differences on the right-hand side of (1.25) are defined. Integrating both sides of (1.25) with respect to v on the interval $[-\delta, 0]$ and dividing by δ, we obtain

$$|\Delta_h^k g(x)| \leqslant \frac{1}{\delta} \sum_{i=0}^{k} \binom{k}{i} \left\{ \int_{-\delta}^0 |\Delta_{i(v+h)/k}^k g[x + (k - i)h]| \, dv \right.$$

$$\left. + \int_{-\delta}^0 |\Delta_{-h+i(v+h)/k}^k g(x + kh)| \, dv \right\} \qquad (1.26)$$

From inequalities (1.24) and (1.26) we can estimate the value of $|\Delta_h^k g(x)|$ for every $x \in [a, b - kh]$. Letting $\Delta_q^k g(\xi) = 0$, if at least one of the points ξ, $\xi + kq$ does not belong to the interval $[a, b]$, then from (1.24) and (1.26) we obtain the following inequality, which is valid for every $x \in [a, b - kh]$:

$$|\Delta_h^k g(x)| \leqslant \frac{1}{\delta} \sum_{i=1}^k \binom{k}{i} \left\{ \int_0^\delta |\Delta_{i(v-h)/k}^k g(x+ih)| dv + \int_0^\delta |\Delta_{h+i(v-h)/k}^k g(x)| dv \right.$$

$$\left. + \int_{-\delta}^0 |\Delta_{i(v+h)/k}^k g[x+(k-i)h]| dv + \int_{-\delta}^0 |\Delta_{-h+i(v+h)/k}^k g(x+kh)| dv \right\}$$

$$(1.27)$$

Letting $x = t + u$ and integrating this inequality with respect to u on the interval $[0, s]$, we obtain

$$\int_0^s |\Delta_h^k g(t+u)| du \leqslant \frac{1}{\delta} \sum_{i=1}^k \binom{k}{i} \left\{ \int_0^\delta \int_0^s |\Delta_{i(v-h)/k}^k g(t+u+ih)| du\, dv \right.$$

$$+ \int_0^\delta \int_0^s |\Delta_{h+i(v-h)/k}^k g(t+u)| du\, dv$$

$$+ \int_{-\delta}^0 \int_0^s |\Delta_{i(v+h)/k}^k g[t+u+(k-i)h]| du\, dv$$

$$\left. + \int_{-\delta}^0 \int_0^s |\Delta_{-h+i(v+h)/k}^k g(t+u+kh)| du\, dv \right\}$$

By changing the variables we obtain:

$$\int_0^s |\Delta_h^k g(t+u)| du \leqslant \frac{1}{\delta} \sum_{i=1}^k \binom{k}{i} \frac{k}{i} \left\{ \left(\int_{-ih/k}^{i(\delta-h)/k} \int_{ih}^{s+ih} + \int_{(k-i)h/k}^{h+i(\delta-h)/k} \int_0^s + \int_{i(h-\delta)/k}^{ih/k} \right. \right.$$

$$\times \int_{(k-i)h}^{s+(k-i)h} + \int_{-h+i(h-\delta)/k}^{-h+ih/k} \int_{kh}^{s+kh} \left) |\Delta_v^k g(t+u)| du\, dv \right\}$$

$$\leqslant \frac{4}{\delta} \sum_{i=1}^k \binom{k}{i} \frac{k}{i} \int_{-\delta}^\delta \int_0^{kh+s} |\Delta_v^k g(t+u)| du\, dv$$

$$\leqslant \frac{(k+1)2^{k+1}}{\delta} \int_{-\delta}^\delta \int_0^{s+k\delta} |\Delta_v^k g(t+u)| du\, dv$$

since

$$\sum_{i=1}^k \binom{k}{i} \frac{k}{i} \leqslant (k+1)2^{k-1} \qquad \blacksquare$$

Proof of Theorem 1.5. We have

$$\omega_k(f, x; \delta) = \sup \left\{ |\Delta_h^k f(t)| : t, t+kh \in \left[x - \frac{k\delta}{2}, x + \frac{k\delta}{2} \right] \cap [a,b], h \geqslant 0 \right\}$$

$$= \sup \left\{ \left| \int_0^h \Delta_h^{k-1} f'(t+u) du \right| : t, t+kh \right.$$

$$\in\left[x-\frac{k\delta}{2},x+\frac{k\delta}{2}\right]\cap[a,b],h\geqslant 0\right\}$$

$$\leqslant\sup\left\{\int_0^{k\delta}\left|\Delta_h^{k-1}f'\left(x-\frac{k\delta}{2}+u\right)\right|du:0\leqslant h\leqslant\delta\right\}$$

From this, using Lemma 1.4 for $k-1$, we obtain

$$\omega_k(f,x;\delta)\leqslant\frac{k2^k}{\delta}\int_{-\delta}^{\delta}\int_0^{(2k-1)\delta}\left|\Delta_v^{k-1}f'\left(x-\frac{k\delta}{2}+u\right)\right|dv\,du$$

and therefore

$$\tau_k(f;\delta)_p=\|\omega_k(f,\cdot;\delta)\|_{L_p}$$

$$\leqslant\frac{k2^k}{\delta}\int_{-\delta}^{\delta}\int_0^{(2k-1)\delta}\left\|\Delta_v^{k-1}f'\left(\cdot-\frac{k\delta}{2}+u\right)\right\|_{L_p}dv\,du$$

$$\leqslant\frac{k2^k}{\delta}\int_{-\delta}^{\delta}\int_0^{(2k-1)\delta}\omega_{k-1}(f';\delta)_p\,dv\,du$$

$$=k(2k-1)2^{k+1}\delta\omega_{k-1}(f';\delta)_p\qquad\blacksquare$$

The constant $c(k)=k(2k-1)2^{k+1}$ is not the best. It is not very difficult to calculate that for $k=2$,

$$\tau_2(f;\delta)_p\leqslant 16\delta\omega(f';\delta)_p\qquad(1.28)$$

i.e. we can take 16 as a constant rather than $c(2)=48$. Even this constant, 16, can be improved.

1.5. Notes

The proof of Marchaud's inequality for the averaged moduli of smoothness is given by Popov [6]. One important generalization of Theorem 1.5 is obtained by Ivanov [6] and is as follows:

Let the function f be continuous. Then

$$\tau_k(f;\delta)_p\leqslant c(k,p)\delta^{1/p}\int_0^\delta\omega_k(f;t)_p t^{-1-1/p}\,dt\qquad(1.29)$$

Many consequences follow from this inequality (see Ivanov [6]).

There exist some generalizations of the averaged moduli of smoothness. First, it is possible to consider averaged moduli of smoothness of fractional order (or index), analogous to the usual moduli of fractional order, given, for example, in Butzer *et al.* [1]. The definition is as follows (see Drianov [2] and Ivanov [4]:

For every $\alpha>0$, $h>0$ let

$$\Delta_h^\alpha f(t)=\sum_{j=0}^\infty\binom{\alpha}{j}(-1)^j f(t-jh)$$

Then the averaged moduli of smoothness of fractional order $\alpha > 0$ are given by

$$\tau_\alpha(f; \delta)_p = \| \omega_\alpha(f, x; \delta) \|_p$$

where

$$\omega_\alpha(f, x; \delta) = \sup \left\{ |\Delta_h^\alpha f(t)| : t, t + \alpha h \in \left[x - \frac{\alpha \delta}{2}, x + \frac{\alpha \delta}{2} \right] \right\}$$

We consider here only the 2π-periodic case.

Let $f \in C$ or L_p and $\int_0^{2\pi} f(x)\, dx = 0$. Weyl introduces the following definition of the α-integral of f, $\alpha > 0$, (see, for example, Zygmund [1]):

$$f_\alpha(x) = \frac{1}{2\pi} \int_0^{2\pi} \psi_\alpha(x - t) f(t)\, dt$$

where

$$\psi_\alpha(t) = \sum_{k=-\infty}^{\infty} e^{ikt}(ik)^{-\alpha}$$

Drianov [2] gives the following results. If $\alpha > \beta > 1/p$ then $(1 \leqslant p \leqslant \infty)$

(1) $\tau_\alpha(f_\alpha; \delta)_p \leqslant c(\alpha, p) \delta^\alpha \| f \|_p$

(2) $\tau_\alpha(f_\alpha; \delta)_p \leqslant c(\alpha, \beta, p) \delta^\beta \| f_{\alpha - \beta} \|_p$

(3) $\tau_\alpha(f_\alpha; \delta)_p \leqslant c(\alpha, \beta, p) \delta^\beta \omega_{\alpha - \beta}(f_{\alpha - \beta}; \delta)_p$

(4) $\tau_\alpha(f_\alpha; \delta)_p \leqslant c(\alpha, \beta, p) \delta^\beta \tau_{\alpha - \beta}(f_{\alpha - \beta}; \delta)_p$

Inequality (2) for $\alpha = 1$ was obtained by Ivanov [4].

An other generalization of the τ-moduli in L_p can be shown by taking the norm of the local modulus, not in L_p, but in some other norm, for example in some Orlić spaces or modular spaces (see Musielak [1] and Taberski [1]).

The definition of τ-moduli for 2π-periodic multivariate functions is given by Popov [8], where some of the properties are stated (for the proofs see Popov and Hristov [1]). In the multivariate case some spatial effects appear. For example property (6) has the following form:

$$\tau_1(f; \delta)_p \leqslant c(m) \sum_{0 \leqslant \alpha_i \leqslant 1, |\alpha| > 0} \delta^{|\alpha|} \| D^\alpha f \|_p$$

where, as usual

$$D^\alpha f = \frac{\partial^{|\alpha|} f}{\partial^{\alpha_1} x_1 \cdots \partial^{\alpha_m} x_m} \qquad \alpha = (\alpha_1, \ldots, \alpha_m), \ |\alpha| = \sum_{i=1}^m \alpha_i$$

Popov [11] shows that for $m = 2$, $p = 1$, the term

$$\delta^2 \left| \frac{\partial^2 f}{\partial x_1 \, \partial x_2} \right|$$

cannot be omitted.

Totkov [2] [3] consider the connection between the τ-moduli and different variations of multivariate functions.

An analogous result to Theorem 1.5 for the multivariate case is given by Hristov [7]. Ivanov [7] generalizes inequality (1.29) for the multivariate case in the following way. If the function f is continuous, then for $k > m/p$ we have

$$\tau_k(f;\delta)_p \leqslant c(k,m,p)\delta^{m/p} \int_0^\delta \omega_k(f;t)_p t^{-1-m/p}\,dt$$

where m is the dimension.

Properties (5) and (5′) of the τ-moduli can be improved not only for $k=1$ (see (5″)), but for $k \geqslant 2$ too. In Hristov and Ivanov [1] it is proved that

$$\tau_k(f;\lambda\delta)_p \leqslant c(k)\lambda^K \tau_K(f;\delta)_p$$

for $\lambda \geqslant 1$, where the constant $c(k)$ depends only on k.

2 INTERPOLATION THEOREMS

In this chapter we shall consider two types of theorems. The first type estimate the deviation between functions from their interpolation and approximation polynomials by means of the moduli of smoothness. The basic result here is Whitney's theorem. The second type are based on the Riesz–Torin theorem. We shall use interpolation results of these two types to obtain interpolation theorems which make use of the averaged moduli of smoothness.

2.1. Whitney's theorem

For every bounded function f the following trivial estimate holds:

$$\omega_k(f;\Delta) \leqslant 2^k \|f\|_{C_\Delta}$$

since

$$|\Delta_h^k f(x)| = \left| \sum_{i=0}^k (-1)^{i+k} \binom{k}{i} f(x+ih) \right|$$

$$\leqslant \sum_{i=0}^k \binom{k}{i} \|f\|_{C_\Delta} = 2^k \|f\|_{C_\Delta}$$

This inequality cannot in general be inverted since there exist functions for which $\omega_k(f;\Delta) = 0$ and $\|f\|_{C_\Delta} \neq 0$. Whitney's theorem gives additional conditions which allow us to invert the above inequality.

2.1.1. Introduction

In 1957, Whitney [1] proved the following theorem, which is now classical in approximation theory and numerical analysis.

Theorem 2.1. *(Whitney [1]) For each integer $n \geqslant 1$, there is a number W_n with the following property. For any interval Δ and for any continuous function f on*

Δ there is a polynomial P of degree at most $n-1$ such that

$$|f(x) - P(x)| \leqslant W_n \omega_n(f; \Delta) \quad x \in \Delta \tag{2.1}$$

where ω_n denotes the nth modulus of continuity of the function f on the interval Δ (see Chapter 1).

One very important problem is what can be said about the size of the constants W_n, in particular whether these constants can be bounded independent of n. For a finite interval Δ, the smallest possible constant W_n in (2.1) is clearly independent of the length of Δ. Therefore we can assume that $\Delta = [0, 1]$. We shall call the constants, W_n, Whitney constants.

One way of estimating the size of the Whitney constants is to construct specific methods of polynomial approximation and estimate the error of approximation in terms of ω_n. For example, when $\Delta = [0, 1]$, we shall consider the polynomial $P := P(f)$ which interpolates f at n equally spaced points:

$$P\left(\frac{v}{n-1}\right) = f\left(\frac{v}{n-1}\right) \quad v = 0, \ldots, n-1,$$

The polynomial P then satisfies (2.1) with some other constant and we let W'_n denote the smallest constant in (2.1) which holds for all f. Similarly, we let W''_n denote the corresponding constant for interpolation at the points $v/(n+1)$, $v = 1, \ldots, n$. Of course we have

$$W_n \leqslant W'_n, \quad W_n \leqslant W''_n \quad n = 1, 2, \ldots \tag{2.2}$$

Whitney [2] has shown that Theorem 2.1 also holds for bounded functions. Ivanov [1] proved Theorem 2.1 for Lebesgue integrable functions defined at each point.

For the practical use of estimates of the form (2.1) it is essential to have good estimates for the Whitney constants. Whitney [1] has given the following estimates:

$$W_1 = W_2 = \tfrac{1}{2}, \quad W'_1 = W'_2 = 1$$
$$\tfrac{8}{15} \leqslant W_3 \leqslant \tfrac{7}{10}, \quad \tfrac{1}{2} \leqslant W_4 \leqslant 3.2425, \quad \tfrac{1}{2} \leqslant W_5 \leqslant 10.4$$
$$\tfrac{16}{15} \leqslant W'_3 \leqslant \tfrac{14}{9}, \quad 1 \leqslant W'_4 \leqslant 3.295\,25, \quad 1 \leqslant W'_5 \leqslant 10.4$$

The first estimates of W_n and W'_n which are valid for all n were very pessimistic: $W'_n = 0 \cdot (n^{2n})$ (Brudni i[1]) and (Sendov [4])

$$W'_n \leqslant (n+1)n^n \tag{2.3}$$

Sendov [4] proposed a numerical method based on linear programming for estimating the Whitney constants. Using this method, he has shown that $W_4 \leqslant 1.26$, $W_5 \leqslant 1.31$, $W_6 \leqslant 1.67$, $W'_4 \leqslant 2.85$, $W'_5 \leqslant 3.46$, $W'_6 \leqslant 5.36$. Estimations of this type for $n > 6$ would need more computer power.

Based on this work, Sendov conjectured that for all n

$$W_n \leqslant 1, \quad W'_n \leqslant 2. \tag{2.4}$$

We shall prove here that the Whitney constants W_n are bounded independent of n (Sendov [5], [6]). The main theorem, which is an improvement of Theorem 2.1, is as follows.

Theorem 2.2. *(Sendov [5], [6]). For any function f integrable on $[0, 1]$ and for each integer $n \geqslant 1$ there is a polynomial P of a degree at most $n - 1$ such that*

$$|f(x) - P(x)| \leqslant 6\omega_n\left(f; \frac{1}{n+1}\right) \quad x \in [0, 1], \tag{2.5}$$

For the polynomial P in (2.5), we may take the polynomial which interpolates f at the points $x_v = v/(n+1)$. This gives

$$W_n \leqslant W_n'' \leqslant 6 \quad n = 1, 2, 3, \dots \tag{2.6}$$

Before proceeding with the proof of Theorem 2.2 we remark that the first advance in improving (2.3) was made by Ivanov and Takev [1] who showed that $W_n = O(n \ln n)$. Binev [5] subsequently improved this to $W_n = O(n)$.

2.1.2.   An identity for integrable functions

Let f be defined for every $x \in [0, 1]$ and integrable on $[0, 1]$. We shall fix an integer $n \geqslant 1$ and introduce the notation

$$h := \frac{1}{n+1}, \quad x = vh + t \quad 0 \leqslant t < h \tag{2.7}$$

where v is an integer. The following operator is very important in the proof:

$$\begin{aligned} \varphi_n(f; x) &= \varphi_n(f; vh + t) \\ &= \frac{(-1)^{(n-v)}}{h\binom{n}{v}} \int_0^h \Delta_y^n f(x - vy) \, dy \end{aligned} \tag{2.8}$$

which is defined for all functions $f \in L_1[0, 1]$.

We have the obvious inequalities for $x \in [vh, (v+1)h]$:

$$|\varphi_n(f; x)| \leqslant \frac{1}{\binom{n}{v}} \omega_n(f; h) \tag{2.9}$$

We shall also use the notation

$$l_{n,v}(x) = \prod_{\substack{j=0 \\ j \neq v}}^n \frac{x - j}{v - j} \quad v = 0, 1, \dots, n \tag{2.10}$$

for the basic Lagrange polynomials for interpolation at $0, 1, \dots, n$. The following representation of a function in terms of the operator (2.8) is valid:

Proposition 2.1.   *If f is defined for every $x \in [0, 1]$ and is integrable on $[0, 1]$, then for each integer $n \geq 1$, we have**

$$f(x) = P_{n-1}(x) + \varphi_n(f; x) + \sum_{j=0}^{n} \frac{1}{h} \int_0^t \varphi_n(f; jh + v) l'_{n,j}\left(\frac{x-v}{h}\right) dv \qquad (2.11)$$

Proof.   Let $x \in [vh, (v-1)h]$. Then

$$\varphi_n(f; x) = f(x) + \frac{(-1)^{n-v}}{h \binom{n}{v}} \int_0^h \sum_{\substack{j=0 \\ j \neq v}}^{n} (-1)^{n-j} \binom{n}{j} f[x + (j-v)v]\, dv$$

$$= f(x) + \frac{(-1)^v}{h \binom{n}{v}} \sum_{\substack{j=0 \\ j \neq v}}^{n} \frac{(-1)^j}{j-v} \binom{n}{j} \int_{vh+t}^{jh+t} f(v)\, dv \qquad (2.12)$$

If we define the functions

$$y_j(t) = \int_0^{jh+t} f(v)\, dv \quad 0 \leqslant t \leqslant h,\ j = 0,\ldots,n \qquad (2.13)$$

then (2.12) may be written in the form of a system of linear non-homogeneous differential equations with constant coefficients.

$$y'_v(t) = \frac{(-1)^{v-1}}{h \binom{n}{v}} \sum_{\substack{j=0 \\ j \neq v}}^{n} \frac{(-1)^j}{j-v} \binom{n}{j} (y_j(t) - y_v(t)) + \varphi_n(f; vh + t) \quad v = 0,\ldots,n \quad (2.14)$$

It is easy to see that the general solution of the homogeneous part of the system (2.14) is

$$y_v(t) = \sum_{\delta=0}^{n} C_s (vh + t)^s \quad v = 0, 1, \ldots, n \qquad (2.15)$$

where $C_0, C_1, \ldots, C_n$ are arbitrary constants.

Indeed, it is enough to show that for every $s = 0, 1, \ldots, n$ the functions $y_v(t) = (vh + t)^s$, $v = 0, 1, 2, \ldots, n$, are solutions of the homogeneous part of (2.14) and this follows from

$$\frac{(-1)^{v-1}}{h \binom{n}{v}} \sum_{\substack{j=0 \\ j \neq v}}^{n} \frac{(-1)^j}{j-v} \binom{n}{j} ((jh + t)^s (vh + t)^s)$$

$$= \frac{(-1)^{v-1}}{h \binom{n}{v}} \sum_{\substack{j=0 \\ j \neq v}}^{n} \frac{(-1)^j}{j-v} \binom{n}{j} (j-v) h \sum_{i=0}^{s-1} (jh + 1)^i (vh + t)^{s-i-1}$$

---

*P_{n-1} is a polynomial of degree at most $n - 1$.

$$= \frac{(-1)^{v-1}}{\binom{n}{v}} \sum_{i=0}^{s-1} (vh+t)^{s-i-1} \sum_{\substack{j=0 \\ j\neq v}}^{n} (-1)^j \binom{n}{j}(jh+t)^i$$

$$= \frac{(-1)^{v-1}}{\binom{n}{v}} \sum_{i=0}^{s-1} (vh+t)^{s-i-1}(-1)^{v-1}\binom{n}{v}(vh+t)^i$$

$$= s(vh+t)^{s-1} = [(vh+t)^s]'$$

To find a solution of the non-homogeneous system (2.14) we shall use the well-known method of variation of parameters and look for a solution of the form

$$\eta_v(t) = \sum_{s=0}^{n} C_s(t)(vh+t)^s \quad v=0,1,2,\ldots,n \tag{2.16}$$

The functions $C_s(t)$, $s=0,1,2,\ldots,n$, have to be found as solutions of the system

$$\sum_{s=0}^{n} C_s'(t)(vh+t)^s = \varphi_n(f;vh+t) \quad v=0,1,2,\ldots,n \tag{2.17}$$

Let $D(t)$ be the determinant

$$D(t):= \| a_{v,s} \|, \ a_{v,s} = (vh+t)^s \quad v,s=0,1,2,\ldots,n$$

and $D_{v,s}(t)$ be the conjugate value of $a_{v,s}$, i.e.

$$\sum_{v=0}^{n} a_{v,s}D_{v,s}(t) = D(t) \quad s=0,1,2,\ldots,n \tag{2.18}$$

Then from (2.17) we have

$$C_s'(t) = \frac{1}{D(t)} \sum_{j=0}^{n} D_{j,s}(t)\varphi_n(f;jh+t) \quad s=0,1,2,\ldots,n$$

or

$$C_s(t) = \int_0^t \frac{1}{D(v)} \sum_{j=0}^{n} D_{j,s}(v)\varphi_n(f;jh+v)\,dv \quad s=0,1,2,\ldots,n \tag{2.19}$$

Using (2.16) and (2.19) we obtain

$$\eta_v(t) = \sum_{s=0}^{n} (vh+t)^s \int_0^t \frac{1}{D(v)} \sum_{j=0}^{n} D_{j,s}(v)\varphi_n(f_ijh+v)\,dv$$

$$= \int_0^t \sum_{j=0}^{n} \varphi_n(f;jh+v)\frac{1}{D(v)} \sum_{s=0}^{n} (vh+t)^s D_{j,s}(v)\,dv \tag{2.20}$$

It is easy to see, considering (2.20), that

$$\frac{1}{D(v)} \sum_{s=0}^{n} (vh+t)^s D_{j,s}(v) = l_{n,j}\left(\frac{vh+t-v}{h}\right) = l_{n,j}\left(\frac{x-v}{h}\right)$$

(see (2.10)). Consequently it follows from (2.20) that

$$\frac{1}{D(v)} \sum_{s=0}^{n} (vh + t)^s D_{j,s}(v) = l_{n,j}\left(\frac{vh + t - v}{h}\right) = l_{n,j}\left(\frac{x - v}{h}\right)$$

(see (2.10)). Consequently it follows from (2.20) that

$$\eta_v(t) = \int_0^t \sum_{j=0}^{n} \varphi_n(f; jh + v) l_{n,j}\left(\frac{vh + t - v}{h}\right) dv \quad v = 0, 1, \ldots, n \qquad (2.21)$$

Hence, the general solution of the non-homogeneous system (2.14) will be

$$y_v(t) = \int_0^{vh+t} f(v) \, dv$$

$$= \sum_{s=0}^{n} C_s(vh + t)^s + \int_0^t \sum_{j=0}^{n} \varphi_n(f; jv + v) l_{n,j}\left(\frac{vh + t - v}{h}\right) dv \quad v = 0, 1, 2, \ldots, n$$
$$(2.22)$$

With one differentiation of (2.22) we obtain the identity

$$f(x) = f(vh + t)$$

$$= \sum_{s=1}^{n} s C_s x^{s-1} + \varphi_n(f; x) + h^{-1} \int_0^t \sum_{j=0}^{n} \varphi_n(f; jh + v) l'_{n,j}\left(\frac{x - v}{h}\right) dv \qquad ■$$

Proposition 2.2. *Let $P_{n-1}^*(f)$ be the interpolation polynomial for f at the points $h, 2h, \ldots, nh$, i.e.*

$$P_{n-1}^*(f; x) = \sum_{j=1}^{n} f(ih) l_{n-1, j-1}\left(\frac{x}{h} - 1\right) \qquad (2.23)$$

Then

$$f(x) - P_{n-1}^*(f; x) = \Delta_h^n f(0) l_{n,0}\left(\frac{x}{h}\right) + \varphi_n(f; x) - \sum_{j=0}^{n} \varphi_n(f; jh) l_{n,j}\left(\frac{x}{h}\right)$$

$$+ h^{-1} \int_0^t \sum_{j=0}^{n} \varphi_n(f; jh + v) l'_{n,j}\left(\frac{x - v}{h}\right) dv \qquad (2.24)$$

Proof. From (2.11), we have

$$f(jh) = P_{n-1}(jh) + \varphi_n(f; jh) \quad j = 0, 1, 2, \ldots, n$$

and using (2.11) we find again that

$$f(x) - \sum_{j=0}^{n} f(jh) l_{n,j}\left(\frac{x}{h}\right) = P_{n-1}(x) + \varphi_n(f; x) + h^{-1} \int_0^t \sum_{j=0}^{n} \varphi_n(f; jh + v) l'_{n,j}\left(\frac{x - v}{h}\right) dv$$

$$- P_{n-1}(x) - \sum_{j=0}^{n} \varphi_k(f; jh) l_{n,j}\left(\frac{x}{h}\right) \qquad (2.25)$$

Alternatively

$$l_{n,j}\left(\frac{x}{h}\right) = l_{n-1,j-1}\left(\frac{x}{h} - 1\right) + (-1)^{n-j}\binom{n}{j}l_{n,0}\left(\frac{x}{h}\right) \tag{2.26}$$

for $j = 1, \ldots, n$. From (2.26) and (2.25) we obtain (2.24). ∎

2.1.3. Proof of theorem 2.2

For the proof of Theorem 2.2. we need two lemmas.

Lemma 2.1. *We have*

$$\gamma_n := \max\left\{\sum_{j=0}^{n}\binom{n}{j}^{-1}|l_{n,j}(x)| : 0 \leqslant x \leqslant 1\right\} = 1 \tag{2.27}$$

Proof. We have $\gamma_n = \max\{\gamma(t) : 0 \leqslant t \leqslant 1\}$, where

$$\gamma(t) := \frac{(1-t)(2-t)\cdots(n-1)}{n!} + \frac{t(1-t)\cdots(n-t)}{n!}\sum_{j=1}^{n}\frac{1}{j-1}$$

The polynomial γ is of a degree n and for the $n+1$ consecutive values $k = 0, 1, 2, \ldots, n$ we have

$$\gamma(k) = \frac{(-1)^k}{\binom{n}{k}}$$

Since these values alternative in sign, $\gamma'(t)$ has no more than one zero in the interval $[0, 1]$. On the other hand, $\gamma'(0) = 0$ and

$$\gamma''(0) = \sum_{j=0}^{n}j^{-2} - \left(\sum_{j=0}^{n}j^{-1}\right)^2 < 0$$

This means that $\gamma(t)$ is monotonically decreasing in $[0, 1]$ and $\gamma(t) \leqslant \gamma(0) = 1$ for $t \in [0, 1]$. ∎

Lemma 2.2. *Let*

$$\mu_{n,v} := \sum_{j=0}^{n}\binom{n}{j}^{-1}\max\{|l_{n,j}(x)| : v \leqslant x \leqslant v+1\} \tag{2.28}$$

Then

$$\mu_{n,v} \leqslant \frac{1 + \sigma_v + \sigma_{v+1}}{\binom{n}{v}} \quad v = 0, 1, 2, \ldots, \frac{n-1}{2} \tag{2.29}$$

where

$$\sigma_v := 1 + \frac{1}{2} + \frac{1}{3} + \cdots + \frac{1}{v} \quad \sigma_0 := 0$$

In particular, for $v = 0$, we have

$$\mu_{n,0} = 1 + \sum_{j=1}^{n} \binom{n}{j}^{-1} \max\{|l_{n,j}(x)| : 0 \leqslant x \leqslant 1\} \leqslant 2 \qquad (2.30)$$

Proof. Let $v \leqslant x \leqslant v + 1$ and $x = v + t$; $0 \leqslant t \leqslant 1$, then, since $v \leqslant [(n-1)/2]$, we obtain

$$\mu_{n,v} = \frac{1}{n!} \sum_{j=0}^{n} \max\left\{\frac{(v+t)\cdots(1+t)t(1-t)\cdots(n-v-t)}{|v-j+t|} : 0 \leqslant t \leqslant 1\right\}$$

$$\leqslant \frac{1}{\binom{n}{v}} + \frac{1}{n!} \sum_{\substack{j=0 \\ j \neq v}}^{n} \max\left\{\left|\frac{1+t/(v-j)}{1-t}\right| \left|\frac{(v+t)\cdots(1+t)t(1-t)\cdots(n-v-t)}{|v-j+t|}\right| : 0 \leqslant t \leqslant 1\right\}$$

$$= \frac{1}{\binom{n}{v}} + \frac{1}{n!} \max\{(v+t)\cdots(1+t)t(2-t)\cdots(n-v-t) : 0 \leqslant t \leqslant 1\} \sum_{\substack{j=0 \\ j \neq v}}^{n} \frac{1}{|v-j|}$$

$$= \frac{1}{\binom{n}{v}} + \frac{1}{\binom{n}{v}} \max\{q_{n,v}(t) : 0 \leqslant t \leqslant 1\} \qquad (2.31)$$

where

$$q_{n,v}(t) := (\sigma_v + \sigma_{n-v})\left(1 + \frac{t}{v}\right)\cdots\left(1 + \frac{t}{1}\right)t\left(1 - \frac{t}{2}\right)\cdots\left(1 - \frac{t}{n-v}\right)$$

We shall show that

$$\max_{0 \leqslant t \leqslant 1} q_{n,v}(t) \leqslant \max_{0 \leqslant t \leqslant 1} q_{2v+1,v}(t)$$

$$= (\sigma_y + \sigma_{v+1}) \max_{0 \leqslant t \leqslant 1} t\left[1 - \frac{t(1-t)}{12}\right]\cdots\left[1 - \frac{t(1-t)}{v(v+1)}\right]$$

$$= \sigma_v + \sigma_{v+1} \qquad (2.32)$$

It is easy to see that

$$q_{n,v}(0) = q_{n-1,v}(0) = 0$$

$$q'_{n,v}(0) = \sigma_v + \sigma_{n-v} > \sigma_v + \sigma_{n-v-1} = q_{n-1,v}(0)$$

$$q_{n,v}(1) = \frac{v+1}{n-v}(\sigma_v + \sigma_{n-v}) < \frac{v+1}{n-v-1}(\sigma_v + \sigma_{n-v-1}) = q_{n-1,v}(1) \qquad (2.33)$$

There exists only one point $\tau'_n \in (0, 1)$ such that

$$q_{n,v}(\tau'_n) = q_{n-1,v}(\tau'_n)$$

and for this point we have,

$$(\sigma_v + \sigma_{n-v})\left(1 - \frac{\tau'_n}{n-v}\right) = \sigma_v + \sigma_{n-v-1}$$

or

$$\tau'_n = \frac{1}{\sigma_v + \sigma_{n-v}}$$

The only zero τ''_n of the polynomial $q'_{n,v}(t)$ in the interval $[0, 1]$ is also a solution of the equation

$$\frac{1}{t} + \sum_{j=1}^{v} \frac{1}{j+t} - \sum_{j=2}^{n-v} \frac{1}{j-t} = 0$$

From this we obtain

$$\frac{1}{\tau''_n} < \sum_{j=2}^{n-v} \frac{1}{j-t} < \sigma_{n-v-1} < \sigma_{n-v}$$

and hence

$$\tau''_n > \frac{1}{\sigma_{n-v}} \geqslant \tau'_n \qquad (2.34)$$

Now (2.32) follows from (2.34) and (2.33), and (2.28) and (2.29) then follow from (2.31) and (2.32). ∎

Proposition 2.3. *For any function f which is integrable on* $[0, 1]$, *we have*

$$\left| f(x) - \sum_{j=1}^{n} f(jh) l_{n-1,j-1}\left(\frac{x}{h} - 1\right) \right| \leqslant \frac{6 + 7 \min(\sigma_v, \sigma_{n-v})}{\binom{n}{v}} \omega_n(f; h) \qquad (2.35)$$

for $x \in [vh, (v+1)h]$, $h := 1/(n+1)$, $v = 0, 1, \ldots, n$, $\delta_v := 1 + \frac{1}{2} + \cdots + 1/v$, $\sigma_0 := 0$.

Proof. Since the knots of the interpolation polynomial

$$P^*_{n-1}(f; x) = \sum_{j=1}^{n} f(jh) l_{n-1,j-1}\left(\frac{x}{h} - 1\right)$$

in (2.35) are symmetric with respect to the middle of the interval $[0, 1]$, it is sufficient to prove (2.35) only for $x \in [0, \frac{1}{2}]$ or for $v = 0, \ldots, [(n-1)/2]$.

For $v = 0$, from Lemmas 2.1 and 2.2, using (2.24) and (2.9), we obtain

$$|f(x) - P^*_{n-1}(f; x)|$$

$$\leqslant \omega_n(f; h)\left[\max_{0 \leqslant t \leqslant 1} |l_{n,0}(t)| + 1 + \max_{0 \leqslant t \leqslant 1} \sum_{j=0}^{n} \frac{1}{\binom{n}{j}} |l_{n,j}(t)| + \sum_{j=0}^{n} \frac{1}{\binom{n}{j}} \int_0^t |l'_{n,j}(v)| \, dv \right]$$

$$\leqslant \omega_n(f, h)\left[3 + 1 + \sum_{j=1}^{n} \frac{1}{\binom{n}{j}} \max_{0 \leqslant t \leqslant 1} |l_{n,j}(t)| \right]$$

$$\leqslant 6\omega_n(f; h)$$

For $v = 1, 2, 3, \ldots, [(n-1)/2]$ we have analogously

$$|f(x) - P_{n-1}^*(f;x)| \leqslant \frac{\omega_n(f;h)}{\binom{n}{v}}\left[2 + \binom{n}{v}\max_{v \leqslant u \leqslant v+1}\sum_{j=0}^{n}\frac{1}{\binom{n}{j}}|l_{n,j}(u)|\right.$$

$$\left. + \binom{n}{v}\sum_{j=0}^{n}\frac{1}{\binom{n}{j}}\int_0^1 |l'_{n,j}(v+v)|\,\mathrm{d}v\right]$$

$$\leqslant \frac{\omega_n(f;h)}{\binom{n}{v}}\left\{2 + 3\binom{n}{v}\sum_{j=0}^{n}\frac{1}{\binom{n}{j}}\max_{v \leqslant u \leqslant v+1}|l_{n,j}(u)|\right\}$$

$$\leqslant \frac{5 + 3(\sigma_v + \sigma_{v+1})}{\binom{n}{v}}\omega_n(f;h) \leqslant \frac{6 + 7\delta_v}{\binom{n}{v}}\omega_n(f;h) \qquad \blacksquare$$

Since for every $v = 0, 1, 2, \ldots, n$ we have

$$\frac{6 + 7\min(\sigma_v, \sigma_{n-v})}{\binom{n}{v}} \leqslant 6,$$

Theorem 2.2 follows immediately from Proposition 2.3.

2.1.4.   Some other results related to Whitney's theorem

Lemma 2.3. $W_2' = 1$.

Proof.   It is sufficient to consider a function f, for which $\|f\|_{M[a,b]} = 1$, $f(a) = f(b) = 0$. Let us assume that $[a, b] = [0, 1]$. Since $\|f\|_{M[0,1]} = 1$, for every $\varepsilon > 0$ there is $x_\varepsilon \in [0, 1]$ such that $|f(x_\varepsilon)| > 1 - \varepsilon$. Let us also assume that $x_\varepsilon \in [0, \frac{1}{2}]$. Then

$$\omega_2(f; [0, 1]) \geqslant |f(0) - 2f(x_\varepsilon) + f(2x_\varepsilon)|$$
$$\geqslant 2|f(x_\varepsilon)| - |f(2x_\varepsilon)|$$
$$\geqslant 2 - 2\varepsilon - 1 = 1 - 2\varepsilon$$

Therefore $\omega_2(f; [0, 1]) \geqslant 1$, i.e. $W_2' \leqslant 1$.
Conversely, for the function

$$f(x) = \begin{cases} 0, & x = 0 \\ 1 - x, & 0 < x \leqslant 1 \end{cases}$$

we have $\|f\|_{M[0,1]} = 1$, $f(0) = f(1) = 0$, $\omega_2(f; [0, 1]) = 1$, therefore $W_2' = 1$. $\blacksquare$

Lemma 2.4.   *Analogously to Lemma 2.3 we can prove that* $W_k' \geqslant 1$, $k = 2, 3, \ldots$.

Proof. Let $Q_k(f)$ be the interpolation polynomial for the function f at the points $v/(k-1)$, $v = 0, \ldots, k-1$. Then for the function

$$f_k(x) = \begin{cases} 0, & x = 0 \\ (1 - (k-1)x) \cdots ((k-1) - (k-1)x)/(k-1), & 0 < x \leqslant 1 \end{cases}$$

we have $\| f_k - Q_k(f_k) \|_{M[0,1]} = 1$, $\Delta_h^k f_k(0) = 1$, $\Delta_h^k f_k(x) = 0$ for $x \in [0,1]$, therefore $\omega_k(f_k; [0,1]) = 1$, $W_k' \geqslant 1$. $\blacksquare$

We shall call the polynomial $p = p_k(f)$, for which Theorem 2.1 is valid, Whitney's polynomial for the function f of a degree $k - 1$.

Let $M[a,b]$ denote the space of all functions which are bounded and measurable on the interval $[a,b]$.

Theorem 2.3. *Let L be a bounded linear operator on $M[a,b]$ and let $L(P) = P$ for every polynomial $P \in H_{k-1}$, where H_{k-1} is the set of all algebraic polynomials of degree $k - 1$. Then for every function $f \in M[a,b]$ we have*

$$\| f - L(f) \|_{C[a,b]} \leqslant (1 + \| L \|_{M[a,b]}) W_k \omega_k \left(f; \frac{b-a}{k} \right)$$

where W_k is Whitney's constant.

Proof. Let $p_k(f)$ be Whitney's polynomial for f of degree $k - 1$. Then, using Whitney's Theorem 2.1 we obtain:

$$\| f - L(f) \|_C \leqslant \| f - p_k(f) \|_C + \| p_k(f) - L(p_k(f)) \|_C + \| L(p_k(f)) - L(f) \|_C$$

$$\leqslant W_k(1 + \| L \|_M)\omega_k(f; [a,b]) = W_k(1 + \| L \|_M)\omega_k \left(f; \frac{b-a}{k} \right) \quad \blacksquare$$

Theorem 2.4. *Let F be a bounded linear functional on $M[a,b]$. Let $F(P) = 0$ for every $P \in H_{k-1}$. Then for every $f \in M[a,b]$ we have*

$$|F(f)| \leqslant W_k \| F \|_{M[a,b]} \omega_k \left(f; \frac{b-a}{k} \right)$$

Remark. By Theorem 2.2, in both theorems we can replace W_k by 6.

2.2. Theorem for intermediate approximations

The following theorem is very useful in many equations connected with the approximation of functions (see Brudnii [1]).

Theorem 2.5. *Let f be a function belonging to $L_p[a,b]$, $1 \leqslant p \leqslant \infty$. For every integer $k > 0$ and every h, $0 < h \leqslant (b-a)/k$, there exists a function $f_{k,h} \in L_p$ with the properties:*

(i) $|f(x) - f_{k,h}(x)| \leqslant c_1(k)\omega_k(f, x; 2h)$

(ii) $\| f - f_{k,h} \|_{L_p[a,b]} \leqslant c_1(k)\omega_k(f;h)_p$

(iii) $f_{k,h} \in W_p^k[a,b]$ (see Notation 6) *and*

$$\| f_{k,h}^{(s)} \|_p \leqslant c_2(k)h^{-s}\omega_s(f;h)_p, \quad s = 1, 2, \ldots, k,$$

where $c_1(k)$ and $c_2(k)$ are constants, depending only on k. We have $c_1(k) = 1$ if the function f is $(b - a)$-periodical.

Remark. Obviously property (i) is defined only for functions bounded on the interval $[a, b]$.

Proof. For simplicity we shall prove Theorem 2.5 only in the case where f is a $(b - a)$-periodic function. Then for every $h > 0$ we can define the function (the modified Steklov's function, see Sendov [7]):

$$f_{k,h}(x) = (-h)^{-k} \int_0^h \cdots \int_0^h \left\{ -f(x + t_1 + \cdots + t_k) + \binom{k}{1} f\left[x + \frac{k-1}{k}(t_1 + \cdots + t_k) \right] \right.$$

$$\left. + \cdots + (-1)^k \binom{k}{k-1} f\left(x + \frac{t_1 + \cdots + t_k}{k} \right) \right\} dt_1 \cdots dt_k \tag{2.36}$$

Obviously we have property (i):

$$|f_{k,h}(x) - f(x)| \leqslant h^k \int_0^h \cdots \int_0^h |\Delta_{(t_1 + \cdots + t_k)/k}^k f(x)| \, dt_1 \cdots dt_k$$

$$\leqslant \omega_k(f, x; 2h)$$

Inequality (ii) is also valid:

$$\| f_{k,h} - f \|_{L_p[a,b]} \leqslant h^{-k} \int_0^h \cdots \int_0^h \| \Delta_{(t_1 + \cdots + t_k)/k}^k f(x) \|_{L_p[a,b]} \, dt_1 \cdots dt_k$$

$$\leqslant \omega_k(f;h)_p$$

For the sth derivative of the function $f_{k,h}$ we have

$$f_{k,h}^{(s)}(x) = (-h)^{-k} \int_0^h \cdots \int_0^h \left\{ -\Delta_h^s f(x + t_1 + \cdots + t_{k-s}) \right.$$

$$+ \binom{k}{1}\binom{k}{k-1}^s \Delta_{(k-1)h/k}^s f\left[x + \frac{k-1}{k}(t_1 + \cdots + t_{k-s}) \right] + \cdots$$

$$\left. + (-1)^k \binom{k}{k-1} k^s \Delta_{h/k}^s f\left(x + \frac{t_1 + \cdots + t_{k-s}}{k} \right) \right\} dt_1 \cdots dt_{k-s} \tag{2.37}$$

for almost all $x \in [a, b]$. Thus we obtain

$$\| f_{k,h}^{(s)} \|_p \leqslant h^{-k} \int_0^h \cdots \int_0^h \left\{ \| \Delta_h^s f(x + t_1 + \cdots + t_{k-s}) \|_p \right.$$

$$\left. + \binom{k}{1}\binom{k}{k-1}^s \left\| \Delta_{(k-1)h/k}^s f\left[x + \frac{k-1}{k}(t_1 + \cdots + t_{k-s}) \right] \right\|_p \right.$$

$$+ \cdots + \binom{k}{k-1} k^s \left\| \Delta^s_{h/k} f\left(x + \frac{t_1 + \cdots + t_{k-s}}{k}\right)\right\|_p \right\} dt_1 \cdots dt_{k-s}$$

$$\leqslant h^{-s} \left\{ \omega_s(f; h)_p + \binom{k}{1}\binom{k}{k-1}^s \omega_s \left[f; \frac{(k-1)h}{k} \right]_p \right.$$

$$\left. + \cdots + \binom{k}{k-1} k^s \omega_s \left(f; \frac{h}{k}\right)_p \right\}$$

$$\leqslant (2k)^k h^{-s} \omega_k(f; h)_p$$

which gives us property (iii) with a constant $c_2(k) = (2k)^k$. ∎

Let us consider the non-periodic case. We cannot use function (2.36), since f is not defined outside the interval $[a, b]$. We shall use as an intermediate approximation the following function, which was introduced by Sendov [6].

$$\hat{f}_{k,h}(x) = (-h)^{-k} \int_0^h \cdots \int_0^h \left\{ -f(x + k\theta_x) + \binom{k}{1} f[x + (k-1)\theta_x] \right.$$

$$\left. + \cdots + (-1)^k \binom{k}{k-1} f(x + \theta_x) \right\} dt_1 \cdots dt_k$$

where

$$\theta_x = \frac{t_1 + \cdots + t_k}{k} - \frac{x-a}{b-a} h \quad 0 \leqslant h \leqslant \frac{b-a}{k}$$

It is clear that with this h for every $x \in [a, b]$ we have

$$a \leqslant x + i\theta_x \leqslant b \quad i = 0, \ldots, k$$

(Using linearity it is sufficient to verify the inequality only for $x = a$ and $x = b$.)
We have therefore

$$|\hat{f}_{k,h}(x) - f(x)| \leqslant \omega_k(f, x; 2h) \tag{2.38}$$

since $|\theta_x| \leqslant h$ for $x \in [a, b]$.
On the other hand

$$\|\hat{f}_{k,h} - f\|_{L_p[a,b]} \leqslant \| \omega_k(f, \cdot; 2h)\|_{L_p[a,b]}$$

$$\leqslant \tau_k(f; 2h)_p \tag{2.39}$$

For the derivatives of the function $\hat{f}_{k,h}$ we have, for $s \leqslant k$,

$$\hat{f}^{(s)}_{k,h}(x) = (-h)^{-k} \int_0^h \cdots \int_0^h \left\{ -\left(1 - \frac{kh}{b-a}\right)^s \Delta^s_h f\left[x - \frac{k(x-a)h}{b-a} + t_1 + \cdots + t_{k-s}\right] \right.$$

$$+ \binom{k}{1}\left(1 - \frac{(k-1)h}{b-a}\right)^s \binom{k}{k-1}^s \Delta^s_{(k-1)h/t} f\left[x - \frac{(k-1)(x-a)h}{b-a}\right.$$

$$\left. + \frac{k-1}{k}(t_1 + \cdots + t_{k-s})\right] + \cdots + (-1)^k \binom{k}{k-1}\left(1 - \frac{h}{b-a}\right)^s$$

$$\left. \times k^s \Delta^s_{h/k} f\left[x - \frac{(x-a)h}{b-a} + \frac{1}{k}(t_1 + \cdots + t_{k-s})\right] \right\} dt_1 \cdots dt_{k-s}$$

Since

$$\left| \Delta^s_{ih/k} f\left[x - \frac{i(x-a)}{b-a} h + \frac{i}{k}(t_1 + \cdots + t_{k-s}) \right] \right| \leq \omega_s(f, x; 2kh) \quad i = 1, 2, \ldots, k$$

we obtain from the above inequality that

$$|\hat{f}^{(s)}_{k,h}(x)| \leq h^{-k} h^{k-s} k^k 2^k \omega_s(f; x; 2kh) \tag{2.40}$$

From inequality (2.40), using the properties of τ_s, we obtain

$$\|\hat{f}^{(s)}_{k,h}\|_p \leq h^{-s} c(k) \tau_s(f; h)_p \tag{2.41}$$

We have obtained inequalities (2.38), (2.39), (2.41) for $0 < h \leq (b-a)/k$. Using the monotonicity of $\omega_k(f, x; h)$ and $\tau_k(f; h)_p$, we obtain for the non-periodic case the following analogue of Theorem 2.5, which will be sufficient for our applications.

Theorem 2.5'. *Let f be a bounded function on the interval $[a, b]$, $f \in L_p[a, b]$, $1 \leq p \leq \infty$. For every integer $k > 0$ and every h, $0 < h \leq (b-a)$, there exists a function $f_{k,h} \in L_p[a, b]$ with the properties:*

(i) $|f(x) - f_{k,h}(x)| \leq \omega_k(f, x; 2h)$
(ii) $\|f - f_{k,h}\|_p \leq \tau_k(f; 2h)_p$
(iii) $f_{k,h} \in W^k_p[a, b]$ *and*

$$\|f^{(s)}_{k,h}\|_p \leq c(t) h^{-s} \tau_s(f; h)_p \quad s = 1, \ldots, k$$

where the constant $c(k)$ is dependent only on k.

2.3. ˌ Riesz–Torin theorem

This important analysis theorem has many applications in different domains. We shall formulate here without proof one variant of this theorem. For the proof see Stein and Weis [1] or Bergh and Löfström [1].

Let T be a linear operator which maps the space $L_p[a, b]$ into the space $L_q[a, b]$ (we assume that the functions are complex valued). If there exists a constant K, for which

$$\|Tf\|_{L_q[a,b]} \leq K \|f\|_{L_p[a,b]}$$

for every function f in $L_p[a, b]$, we say that the operator T is of the type (p, q). The smallest number k with this property is called the (p, q)-norm of the operator T.

Theorem 2.6. *Let T be a linear operator of the type (p_i, q_i) with the (p_i, q_i)-norm K_i, $i = 0, 1$. Then T is an operator of the type (p_t, q_t) with the (p_t, q_t)-norm*

$$K_t \leq K_0^{1-t} K_1^t \tag{2.42}$$

where $1 \leqslant p_i \leqslant \infty$, $1 \leqslant q_i \leqslant \infty$, $i = 0, 1$, $0 < t < 1$,

$$\frac{1}{p_t} = \frac{1-t}{p_0} + \frac{t}{p_1}, \quad \frac{1}{q_t} = \frac{1-t}{q_0} + \frac{t}{q_1}$$

Theorem 2.7. *With the conditions of Theorem 2.6, if the spaces $L_p[a, b]$ are for real-valued functions, the constant K_t in the inequality (2.42) must be replaced by $CK_0^{1-t}K_1^t$.*

We shall use the Riesz–Torin theorem in the following variant.

Theorem 2.8. *Let T be a linear operator, defined in $W_p^r[a, b]$ for $p = 1, \infty, r = 1$. Then for $f_p \in W_p^r$, $p = 1, \infty$, we have*

$$\| Tf_p \|_{L_p} \leqslant M_p \| f_p^{(r)} \|_{L_p} \quad p = 1, \infty$$

Further, if $f \in W_p^r$, $1 < p < \infty$, we have

$$\| Tf \|_{L_p} \leqslant CM_\infty^{1-1/p} M_1^{1/p} \| f^{(r)} \|_{L_p}$$

where C is the constant from Theorem 2.7.

Proof. For every integrable function g we define the linear operator

$$S_g = T\left[\int_a^x \cdots \int_a^{t_3} \int_a^{t_2} g(t_1) \, dt_1 \, dt_2 \cdots dt_r \right]$$

Obviously the operator S satisfies the Riesz–Torin theorem with $p_0 = q_0 = 1$, $p_1 = q_1 = \infty$, since

$$\| Sg \|_{L_p} \leqslant M_p \| g \|_{L_p}$$

where $p = 1, \infty$ for functions g belonging to L_1, L_∞ respectively (by the conditions of the theorem). Therefore, if we take $t = 1 - 1/p$, then $p_t = p$ and for $g \in L_p$ we have

$$\| Sg \|_{L_p} \leqslant CM_1^{1/p} M_\infty^{1-1/p} \| g \|_{L_p} \quad 1 < p < \infty$$

By the condition of the theorem it follows, that $Tq = 0$ for every polynomial q of degree $r - 1$. If $f \in W_p^r$, then

$$f(x) = \int_a^x \cdots \int_a^{t_2} f^{(r)}(t_1) \, dt_1 \cdots dt_r + q(x)$$

where q is a polynomial of degree $r - 1$. Therefore

$$\| Tf \|_{L_p} \leqslant \| Sf^{(r)} \|_{L_p} \leqslant CM_1^{1/p} M_\infty^{1-1/p} \| f^{(r)} \|_{L_p} \qquad \blacksquare$$

2.4. The interpolation theorem for the averaged moduli of smoothness

Lemma 2.5. *Let $\Sigma_n = \{x_i : a = x_0 < \cdots < x_{n+1} = b\}$ be a partition of the interval $[a, b]$ into $n + 1$ subintervals and let $r \geqslant 1$ be an integer. Using the notation*

$\Delta_i = x_{i+1} - x_{i-1}$, $i = 1, 2, \ldots, n$, $d_n = \max\{\Delta_i : 1 \leqslant i \leqslant n\}$, then

$$\left\{ \sum_{i=1}^{n} [\omega_r(f, x_i; 2h)]^p \Delta_i \right\}^{1/p} \leqslant 2^{1/p + 2(r+1)} \tau_r \left(f; h + \frac{d_n}{r} \right)_p,$$

Proof. We have

$$\left\{ \sum_{i=1}^{n} [\omega_r(f, x_i; 2h)]^p \Delta_i \right\}^{1/p} = \left\{ \sum_{i=1}^{n} \int_{x_{i-1}}^{x_{i+1}} [\omega_r(f, x_i; 2h)]^p \, dx \right\}^{1/p}$$

$$\leqslant \left\{ \sum_{i=1}^{n} \int_{x_{i-1}}^{x_{i+1}} \left[\omega_r \left(f, x; 2\left(h + \frac{d_n}{r} \right) \right) \right]^p dx \right\}^{1/p}$$

$$\leqslant 2^{1/p} \tau_r \left(f; 2\left(h + \frac{dn}{r} \right) \right)_p.$$

From these inequalities we obtain

$$\left\{ \sum_{i=1}^{n} [\omega_r(f, x_i; 2h)]^p \Delta_i \right\}^{1/p} \leqslant 2^{1/p + 2(r+1)} \tau_r \left(f; h + \frac{d_n}{r} \right)_p \qquad \blacksquare$$

Definition 2.3. Let $\Sigma_n = \{x_i : a = x_0 < \cdots < x_{n+1} = b\}$ be a partition of the interval $[a, b]$ into $n + 1$ subintervals, $\Delta_i = x_{i+1} - x_{i-1}$, d_n-max Δ_i. The number

$$\| f \|_{l^p \Sigma_n} = \left[\sum_{i=1}^{n} |f(x_i)|^p \Delta_i \right]^{1/p}$$

is called the discrete l^p norm of the function f on the network Σ_n.

The following interpolation theorem is a basic one.

Theorem 2.9. *Let L_n be a linear operator with the properties:*

(a) *If $f \in M[a, b]$ (where $M[a, b]$ is the space of all measurable functions bounded on the interval $[a, b]$), then $L_n(f) \in L_p[a, b]$ $1 \leqslant p \leqslant \infty$, and*

$$\| L_n(f) \|_{L_p} \leqslant K \| f \|_{l^p \Sigma_n}$$

where K is an absolute constant.

(b) *If $f \in W_p^r[a, b]$, then*

$$\| L_n f - f \|_{L_p} \leqslant K_r d_n^s \| f^{(r)} \|_{L_p} \qquad s \leqslant r$$

where K_r is a constant dependent only on r.

Then for every $f \in M[a, b]$ and for $d_n \leqslant \min\{1, ((b - a)/r)^{r/s}\}$ we have

$$\| f - L_n(f) \|_{L_p[a,b]} \leqslant c \tau_r (f; d_n^{s/r})_p$$

where the constant c is dependent only on r, K and K_r.

Remark. Here and in what follows we shall separate the bounded measurable function f from its equivalence class in L_p, i.e. we shall assume that f is given by its values at each point $x \in [a, b]$.

Proof. Let us take $h = (b - a)/r$ and construct for the function f the function $f_{r,h}$ from Theorem 2.5′. We have

$$\|f - L_n(f)\|_p \leqslant \|f - f_{r,h}\|_p + \|f_{r,h} - L_n(f_{r,h})\|_p + \|L_n(f_{r,h}) - L_n(f)\|_p \quad (2.43)$$

From Theorem 2.5′ we obtain

$$\|f - f_{r,h}\|_p \leqslant \tau_r(f; 2h)_p \quad (2.44)$$

Since $f_{r,h} \in W_p^r$ and

$$\|f_{r,h}^{(r)}\|_p \leqslant c(r)h^{-r}\tau_r(f; h)_p$$

from property (b) of the operator L_n we obtain

$$\|L_n(f_{r,h}) - f_{r,h}\|_p \leqslant K_r((r)d_n^s h^{-r}\tau_r(f; h)_p \quad (2.45)$$

From the linearity of L_n and property (a) it follows that

$$\|L_n(f_{r,h}) - L_n(f)\|_p = \|L_n(f_{r,h} - f)\|_p \leqslant K\left\{\sum_{i=1}^n |f(x_i) - f_{r,h}(x_i)|^p \Delta_i\right\}^{1/p} \quad (2.46)$$

Property (i) of $f_{r,h}$ yields

$$|f(x_i) - f_{r,h}(x_1)| \leqslant \omega_r(f, x_i; 2h)$$

which together with (2.46) and Lemma 2.5 gives us

$$\|L_n(f_{r,h}) - L_n(f)\|_p \leqslant K2^{1/p + 2(r+1)}\tau_r\left(f; h + \frac{d_n}{r}\right)_p \quad (2.47)$$

Let us set $h = d_n^{s/r}$. Since $d_n \leqslant 1$, we have $d_n^{s/r} \geqslant h$ ($s \leqslant r$). Therefore from (2.43), (2.44), (2.45) and (2.47), replacing h by $d_n^{s/r}$ ($d_n^{s/r} \geqslant h$), we obtain

$$\|f - L_n f\|_p \leqslant \tau_r(f; 2d_n^{s/r})_p + K_r c(\tau)d_n^s d_n^{-(s/r)r}\tau_r(f; d_n^{s/r})_p$$

$$+ K2^{1/p + 2(\tau+1)}\tau_r\left(f; d_n^{s/r} + \frac{d_n}{r}\right)_p \leqslant c(r, K, K_r)\tau_r(f; d_n^{s/r})_p.$$

Notice that the method of proof gives a large estimate for the constant $c(r, K, K_r)$. ∎

2.5. Notes

Recently Sendov replaced the constant 6 in Theorem 2.2 by 3 (unpublished). The problem of finding the exact Whitney constant remains unsolved. Also an open problem is the constant $c_2(k)$ in Theorem 2.3. It is not known whether or not this constant is bounded. This is obviously relevant to applications.

It is now well known that the usual moduli of smoothness are connected with the K-functional of Peetre. First we recall the definition. If X_0 and X_1 are two Banach (or quasi-Banach) spaces, with $X_1 \subset X_0$, then for every $f \in X_0$ we define the K-functional:

$$K(f,t;X_0,X_1)=\inf\{\|f_0\|_{X_0}+t\|f_1\|_{X_1}:f=f_0+f_1,f_1\in X_1\}$$

The K-functional plays an important role in the theory of interpolation spaces (see Bergh and Löfström [1]).

If we let $X_0=L_p(0,2\pi)$ and for X_1 let the W_p^k be the Sobolev space with the seminorm $\|g\|_{W_p^k}=\|g^{(k)}\|_p$, then we have the following well-known equivalence (for 2π-periodic functions):

$$c_1(k)\omega_k(f;t)_p\leqslant K(f,t^k,L_p,W_p^{(k)})\leqslant c_2(k)\omega_k(f;t)_p\quad 1\leqslant p\leqslant\infty$$

where $c_1(k)$ and $c_2(k)$ are constants dependent only on k.

It is interesting that the averaged moduli are also connected with another type of functional – the so-called one-sided K-functional (Popov [10], [12]), defined as follows: again let X_0 and X_1 be two quasi-Banach spaces of functions $(X_1\subset X_0)$. Let

$$K_+(f;t)\equiv K_+(f,t;X_0,X_1)=\inf\{\|f_0\|_{X_0}+t\|f_1\|_{X_1}:f=f_0+f_1,f_0\geqslant0\}$$
$$K_-(f;t)\equiv K_-(f,t;X_0,X_1)=\inf\{\|f_0\|_{X_0}+t\|f_1\|_{X_1}:f=f_0+f_1,f_0\leqslant0\}$$
$$\tilde{K}(f,t;X_0,X_1)=\max\{K_+(f;t),K_-(f;t)\}$$

Again we set $X_0=L_p(0,2\pi)$, $X_1=W_p^k$ with $\|g\|_{W_p^k}=\|g^{(k)}\|_p$ and we have (Popov [10])

$$c_1(k)\tau_k(f;t)_p\leqslant\tilde{K}(f,t^k;L_p,W_p^k)\leqslant c_2(k)\tau_k(f;t)_p\quad 1\leqslant p\leqslant\infty,$$

where the constant $c_i(k)$, $i=1,2$, is dependent only on k.

Connected with the usual moduli of smoothness are Besov spaces $B_{p,q}^\alpha$, which play an important role in many problems of analysis and differential equations. Norms in Besov spaces $B_{p,q}^\alpha$ are given by

$$\|f\|_{B_{p,q}^\alpha}=\|f\|_p+\left\{\int_0^\infty[t^{-\alpha}\omega_k(f;t)_p]^q\frac{dt}{t}\right\}^{1/q}\quad k>\alpha$$

Using the averaged moduli of smoothness we introduce another space $A_{p,q}^\alpha$ with the following norm:

$$\|f\|_{A_{p,q}^\alpha}=\|f\|_p+\left\{\int_0^\infty[t^{-\alpha}\tau_k(f;t)_p]^q\frac{dt}{t}\right\}^{1/q}\quad k>\alpha$$

(Popov [9], [10], [12]).

Besov spaces are connected with best trigonometric approximations. The spaces $A_{p,q}^\alpha$ are connected with best one-sided approximation (see the notes to Chapter 8).

It is interesting that when the smoothness α is sufficiently large, the Besov spaces coincide with the spaces $A_{p,q}^\alpha$. More precisely, in the one-dimensional case, $B_{p,q}^\alpha=A_{p,q}^\alpha$ (with equivalent norms) if $\alpha>1/p$ (Popov [9], [10], Ivanov [6]). When $\alpha<1/p$, the spaces $B_{p,q}^\alpha$ and $A_{p,q}^\alpha$ are different.

In the multivariate case this problem was considered by Hristov [8] and Ivanov [7]. The corresponding result is as follows.

We have $B^\alpha_{p,q} = A^\alpha_{p,q}$ (with equivalent norms) if $\alpha > m/p$, where m is the dimension, $1 \leqslant p \leqslant \infty$.

The embedding theorems for the spaces $A^\alpha_{p,q}$ are given in the multivariate case by Hristov [6].

The problem of the traces of the functions which belong to the spaces $A^\alpha_{p,q}$ was studied by Alexandrov [3]. With some natural restrictions the trace of a function from the space $A^\alpha_{p,q}$ belongs to the corresponding A-space.

3 NUMERICAL INTEGRATION

In this chapter we shall use the averaged moduli of smoothness (τ-moduli) to estimate the error in the numerical evaluation of definite integrals by means of different quadrature formulae. This new method of estimation means that it is not necessary to make additional assumptions about the integrating function except those which are necessary for applying the corresponding quadrature formula. From these estimates, using the properties of the averaged moduli of smoothness, the orders of the well-known classical estimates for the error of the corresponding quadrature formulae follow for given classes of functions.

In Section 3.1 we consider the simplest quadrature formulae: the rectangle formula, the trapezoidal rule and Simpson's rule. In these elementary cases we clarify the basic idea of the new method of estimating the error using the averaged moduli of the integrand.

In Section 3.2 we give a general method for estimating the error of a composite quadrature formula, in which only the values of the integrand appear. In Section 3.2 we apply this method to some concrete quadrature formulae. We consider also the problem of finding the exact constants in the estimates obtained.

In Section 3.3 general estimations are given for quadrature formulae, which use also the values of the derivatives to a given order. In Section 3.3 these general estimates are applied to concrete quadrature formulae. Quadrature formulae for 2π-periodic functions are considered in Section 3.4. In Section 3.2.2 of this chapter, some problems connected with the optimal quadrature formulae for given classes of functions are considered.

3.1. Estimates for the error of some elementary quadrature formulae

3.1.1. Rectangle formula

The well-known rectangle formula is often used and is the simplest formula for numerical integration.

$$\int_a^b f(x)\,dx = \frac{b-a}{n}\sum_{i=1}^n f(x_i) + R_n^0(f) \tag{3.1}$$

where $x_i = a + (b-a)(2i-1)/2n$, $i = 1, 2, \ldots, n$, are the knots of the quadrature formula, and $R_n^0(f)$ is the error. For the numerical evaluation of the definite integral we take the sum on the right-hand side:

$$\int_a^b f(x)\,dx \approx \frac{b-a}{n}\sum_{i=1}^n f(x_i) \tag{3.2}$$

In numerical analysis texts a representation of the error $R_n^0(f)$ is given which require the additional condition that the function f has a second derivative which is continuous in the interval $[a,b]$. This representation is as follows:

$$R_n^0(f) = \frac{(b-a)^3}{24} f''(\xi) n^{-2} \quad \xi \in [a, b] \tag{3.3}$$

From (3.3) the estimate follows:

$$|R_n^0(f)| \leqslant \frac{(b-a)^3}{24} M_2 n^{-2} \tag{3.4}$$

where $M_2 = \| f'' \|_{C[a,b]}$.

Our aim is to find an estimate for the error $R_n^0(f)$ without the additional condition that the function f should have a continuous second derivative. We shall assume only that the function f is bounded and measurable.

For the error $R_n^0(f)$ we have

$$R_n^0(f) = \int_a^b f(x)\,dx - \frac{b-a}{n}\sum_{i=1}^n f(x_i)$$

$$= \sum_{i=1}^n \left[\int_{\xi_{i-1}}^{\xi_i} f(x)\,dx - h f(x_i) \right]$$

$$= \sum_{i=1}^n \int_0^{h/2} [f(x_i + t) - 2f(x_i) + f(x_i - t)]\,dt,$$

where $\xi_i = a + ih$, $i = 0, 1, \ldots, n$, $h = (b-a)/n$.

Therefore

$$|R_n^0(f)| \leqslant \sum_{i=1}^n \frac{h}{2}\omega_2\left(f, x_i; \frac{h}{2}\right) = \tfrac{1}{2}\sum_{i=1}^n \int_{\xi_{i-1}}^{\xi_i} \omega_2\left(f, x_i; \frac{h}{2}\right) dx$$

$$\leqslant \tfrac{1}{2}\sum_{i=1}^n \int_{\xi_{i-1}}^{\xi_i} \omega_2(f, x; h)\,dx = \frac{1}{2}\tau_2(f; h)_L$$

since for $x \in [\xi_{i-1}, \xi_i]$ we have

$$\omega_2\left(f, x_i; \frac{h}{2}\right) \leqslant \omega_2(f, x; h)$$

We have thus obtained

$$|R_n^0(f)| \leqslant \frac{b-a}{2}\tau_2\left(f;\frac{b-a}{n}\right)_L \tag{3.5}$$

If we work more precisely, the estimation (3.5) can be improved. In fact, for $|t| \leqslant h/2$ we have

$$|f(x_i+t)-2f(x_i)+f(x_i-t)| \leqslant \omega_2\left(f,x_i-t+\frac{h}{2};\frac{h}{2}\right) \tag{3.6}$$

$$|f(x_i+t)-2f(x_i)+f(x_i-t)| \leqslant \omega_2\left(f,x_i+t-\frac{h}{2};\frac{h}{2}\right) \tag{3.7}$$

Using (3.6), we obtain

$$|R_n^0(f)| \leqslant \left|\sum_{i=1}^{n}\int_0^{h/2}[f(x_i+t)-2f(x_i)+f(x_i-t)]\,dt\right|$$

$$\leqslant \sum_{i=1}^{n}\int_0^{h/2}\omega_2\left(f,x_i-t+\frac{h}{2};\frac{h}{2}\right)dt$$

$$= \sum_{i=1}^{n}\int_{x_i}^{\xi_i}\omega_2\left(f,x;\frac{h}{2}\right)dx \tag{3.8}$$

where in the last inequality we have made a change of variables $x_i-t+h/2=x$. Analogously, using (3.7) we obtain

$$|R_n^0(f)| \leqslant \sum_{i=1}^{n}\int_{\xi_{i-1}}^{x_i}\omega_2\left(f,x;\frac{h}{2}\right)dx \tag{3.9}$$

Combining (3.8) and (3.9) we find that

$$2|R_n^0(f)| \leqslant \sum_{i=1}^{n}\int_{\xi_{i-1}}^{\xi_i}\omega_2\left(f,x;\frac{h}{2}\right)dx = \tau_2\left(f;\frac{h}{2}\right)_L$$

So we have obtained the following result.

Theorem 3.1. *For the rectangle formula (3.1), the following estimate for the error holds:*

$$|R_n^0(f)| \leqslant \frac{1}{2}\tau_2\left(f;\frac{(b-a)}{2n}\right)_L$$

Using properties (4) and (6) from Section 1.3, we obtain immediately that

$$\tau_2\left(f;\frac{h}{2}\right)_L \leqslant \frac{(b-a)^3}{4n^2}\|f''\|_C$$

From this and from Theorem 3.1 we obtain estimate (3.4) with accuracy to a factor of $\frac{1}{6}$. However, from Theorem 3.1 we can deduce many other properties using the properties of the averaged moduli. First, we find that if f is Riemann integrable, then $R_n^0(f) = o(1)$. This is an elementary fact of analysis, but it does not follow from estimates (3.3) and (3.4). If the function f is absolutely continuous in the interval $[a, b]$, then using Theorem 3.1 and inequality (1.28) we obtain the estimate

$$|R_n^0(f)| \leq 4 \frac{(b-a)}{n} \omega_1\left(f'; \frac{b-a}{2n}\right)_L \tag{3.10}$$

There are some interesting consequences of (3.10). First, if f is an absolutely continuous function on the interval $[a, b]$, then $R_n^0(f) = o(n^{-1})$, since for every integrable function g we have

$$\lim_{\delta \to 0} \omega(g; \delta)_L = 0$$

If f' has bounded variation, then property (7), Section 1.3, gives $\omega(f'; \delta)_L = O(\delta)$ and therefore

$$R_n^0(f) = O(n^{-2})$$

Since τ_2 can be estimated by τ_1 (see property (3), Section 1.3), we see from Theorem 3.1 and property (7), Section 1.3, that if f has bounded variation, then $R_n^0(f) = O(n^{-1})$. We summarize the basic corollaries from Theorem 3.1 in the following table

f	$R_n^0(f)$
Riemann integrable	$o(1)$
bounded variation	$O(n^{-1})$
absolutely continuous	$o(n^{-1})$
f' has bounded variation	$O(n^{-2})$

From this table we see that the maximal order $O(n^{-2})$ is obtained from the weaker assumption on f with respect to (3.3) and (3.4); regarding the boundedness of f'' it is sufficient only that f' should have bounded variation. We shall see that this effect is typical of all estimates using the averaged moduli of smoothness.

3.1.2. Trapezoidal rule

The composite trapezoidal rule is obtained by dividing the interval $[a, b]$ into n equal subintervals using the points $x_i = a + ih$, $i = 0, 1, \ldots, n$, $h = (b-a)/n$. Let

$$\int_a^b f(x)\,dx \approx \frac{b-a}{2n}\left[f(a) + 2 \sum_{i=1}^{n-1} f(x_i) + f(b) \right] \tag{3.11}$$

The usual estimate of the error

$$R_n^1(f) = \int_a^b f(x)\,dx - \frac{b-a}{2n}\left[f(a) + 2 \sum_{i=1}^{n} f(x_i) + f(b) \right] \tag{3.12}$$

needs a bounded second derivative f'' and is

$$|R_n^1(f)| \leqslant \frac{M_2}{12n^2}(b-a)^3 \quad M_2 = \|f''\|_{C[a,b]} \tag{3.13}$$

The estimate for $|R_n^1(f)|$ by $\tau_2(f;h)_L$ can be obtained in the following way (compare with the estimate for $R_n^0(f)$):

$$R_n^1(f) = \int_a^b f(x)\,dx - \frac{b-a}{2n}\left[f(x_0) + 2\sum_{i=1}^{n-1}f(x_i) - f(x_n)\right]$$

$$= \sum_{i=1}^n \int_{x_{i-1}}^{x_i}\left[f(x) - \frac{f(x_i)+f(x_{i-1})}{2}\right]dx$$

$$= \sum_{i=1}^n \left\{\int_{x_{i-1}}^{x_i}\left[f(x) - f\left(\frac{x_i+x_{i-1}}{2}\right)\right]dx\right.$$

$$\left. -\tfrac{1}{2}\int_{x_{i-1}}^{x_i}\left[f(x_i) - 2f\left(\frac{x_i+x_{i-1}}{2}\right) + f(x_{i-1})\right]dx\right\}$$

from which

$$|R_n^1(f)| \leqslant \sum_{i=1}^n \tfrac{1}{2}\int_{x_{i-1}}^{x_i} \omega_2\left(f, \frac{x_i+x_{i-1}}{2}; \frac{h}{2}\right)dx + \frac{1}{2}\tau_2\left(f; \frac{h}{2}\right)_L$$

$$\leqslant \tfrac{1}{2}\sum_{i=1}^n \int_{x_{i-1}}^{x_i} \omega_2(f,x;h)\,dx + \frac{1}{2}\tau_2\left(f; \frac{h}{2}\right)_L = \tau_2(f;h)_L$$

We have used Theorem 3.1 and the fact that for $x \in [x_{i-1}, x_i]$

$$\omega_2\left(f, \frac{x_i+x_{i-1}}{2}; \frac{h}{2}\right) \leqslant \omega_2(f,x;h)$$

We thus obtain the following result.

Theorem 3.2. *For the error $R_n^1(f)$ of the composite trapezoidal rule (3.11) we have the following estimate:*

$$|R_n^1(f)| \leqslant \tau_2\left(f; \frac{b-a}{n}\right)_L$$

From this theorem, using the properties of the averaged moduli from Section 1.3, we obtain, in the same way as from Theorem 3.1, the results given in the table below:

f	$R_n^1(f)$
Riemann integrable	$o(1)$
bounded variation	$O(n^{-1})$
absolutely continuous	$o(n^{-1})$
f' has bounded variation	$O(n^{-2})$

Note that from Theorem 3.2 we can also obtain estimate (3.13), but with a *worse* constant.

3.1.3. Simpson's rule

Let us now consider the composite Simpson's rule:

$$\int_a^b f(x)\,dx \approx \frac{b-a}{6n}\left[f(a) + 2\sum_{i=1}^{n-1} f(x_i) + 4\sum_{i=1}^{n} f\left(\frac{x_i+x_{i-1}}{2}\right) + f(b) \right] \equiv L_n(f)$$

where $x_i = a + ih$, $i = 0, 1, \ldots, n$, $h = (b-a)/n$. For the error

$$R_n^2(f) = \int_a^b f(x)\,dx - L_n(f) \tag{3.14}$$

with the assumption that f has a fourth bounded derivative, the following estimate is well known:

$$|R_n^2(f)| \leqslant \frac{M_4(b-a)^5}{2880n^4} \qquad M_4 = \| f^{(4)} \|_{C[a,b]} \tag{3.15}$$

We shall estimate $R_n^2(f)$ by $\tau_4(f;h)_L$. It is easy to verify that

$$R_n^2(f) = \int_a^b f(x)\,dx - L_n(f)$$

$$= \sum_{i=1}^{n} \int_{x_{i-1}}^{x_i} \left\{ f(x) - \frac{1}{6}\left[f(x_{i-1}) + 4f\left(\frac{x_i+x_{i-1}}{2}\right) + f(x_i) \right] \right\} dx$$

$$= \sum_{i=1}^{n} \int_{x_{i-1}}^{x_i} [f(x) - P_{3,i}(x)]\,dx \tag{3.16}$$

where $P_{3,i}$ is the Hermite interpolation polynomial of third degree, which interpolates the function f at the points x_{i-1}, $(x_{i-1}+x_i)/2$, x_i, and satisfies the condition

$$P'_{3,i}\left(\frac{x_{i-1}+x_i}{2}\right) = 0$$

Simpson's formula

$$\int_{x_{i-1}}^{x_i} f(x)\,dx = \int_{x_{i-1}}^{x_i} P_{3,i}(x)\,dx + R_{2,i}(f)$$

$$= \frac{1}{6}(x_i - x_{i-1})\left[f(x_{i-1}) + 4f\left(\frac{x_i+x_{i-1}}{2}\right) + f(x_i) \right] + R_{2,i}(f)$$

is exact for every function f that is an algebraic polynomial of degree 3, i.e. $R_n^2(f) = 0$ if $f \in H_3$. Therefore the linear functional

$$L_{3,i}(f) = \int_{x_{i-1}}^{x_i} [f(x) - P_{3,i}(x)]\,dx$$

which is defined for every bounded and integrable function f on the interval $[x_{i-1}, x_i]$ vanishes for every $f \in H_3$. Using Theorem 2.4 we obtain the estimate

$$|L_{3,i}(f)| \leqslant \|L_{3,i}\|_C W_4\left(\dot{}f, \frac{x_i + x_{i-1}}{2}; \frac{b-a}{4n}\right) \tag{3.17}$$

since in $[x_{i-1}, x_i]$ we have

$$\omega_4\left(f; \frac{x_i - x_{i-1}}{4n}\right)_C = \omega_4\left(f, \frac{x_i + x_{i-1}}{2}; \frac{b-a}{4n}\right).$$

It is easy to see that

$$\|L_{3,i}(f)\|_C \leqslant (x_i - x_{i-1})\|f\|_C + \tfrac{1}{6}(x_i - x_{i-1})(\|f\|_C + 4\|f\|_C + \|f\|_C)$$

i.e.

$$\|L_{3,i}\|_C \leqslant 2(x_i - x_{i-1})$$

and therefore we obtain from inequality (3.17)

$$|L_{3,i}(f)| \leqslant 2(x_i - x_{i-1})W_4\omega_4\left(f, \frac{x_1 + x_{i-1}}{2}; \frac{b-a}{4n}\right) \tag{3.18}$$

From (3.16) it follows that

$$|R_n^2(f)| \leqslant \sum_{i=1}^{n} |L_{3,i}(f)| \tag{3.19}$$

Since for every $x \in [x_{i-1}, x_i]$ we have

$$\omega_4\left(f, \frac{x_i + x_{i-1}}{2}; \frac{b-a}{4n}\right) \leqslant \omega_4\left(f, x; \frac{b-a}{2n}\right)$$

we obtain from (3.18) and (3,19) the following estimate:

$$|R_n^2(f)| \leqslant 2W_4 \sum_{i=1}^{n} \int_{x_{i-1}}^{x_i} \omega_4\left(f, \frac{x_i + x_{i-1}}{2}; \frac{b-a}{4n}\right) dx$$

$$\leqslant 2W_4 \sum_{i=1}^{n} \int_{x_{i-1}}^{x_i} \omega_4\left(f, x; \frac{b-a}{2n}\right) dx$$

$$= 2W_4 \tau_4\left(f; \frac{b-a}{2n}\right)_L$$

So we have proved the following theorem.

Theorem 3.3.   *For the error of the composite Simpson's rule the following estimate holds:*

$$|R_n^2(f)| \leqslant 2W_4 \tau_4\left(f; \frac{b-a}{2n}\right)_L$$

where W_4 is Whitney's approximation constant. (see Chapter 2).

From Theorem 3.3, again using the properties of the averaged a moduli from Chapter 1, we obtain the statements in the table below.

f	$R_n^2(f)$
Riemann integrable	$o(1)$
bounded variation	$O(n^{-1})$
$f^{(i)}$, $i = 0, 1, 2, 3$ is absolutely continuous	$o(n^{-i-1})$
$f^{(i)}$, $i = 1, 2, 3$ has bounded variation	$O(n^{-i-1})$

We therefore obtain the order of the classical estimate (3.15) under the weaker restriction: f''' has bounded variation on $[a, b]$. Note also that the constant before the order n^{-4}, which we obtain by this method, is larger than the classical constant $1/2880$.

3.2. Estimation of the error of composite quadrature formulae, using only the values of the integrand

3.2.1. General estimate

Let Lf be a quadrature formula in the interval $[0, 1]$, which uses only the values of the function f, i.e. let

$$Lf = \sum_{i=1}^{m} A_i f(x_i) \quad x_i \in [0, 1], \quad i = 1, 2, \ldots, m$$

$$0 \leqslant x_1 < x_2 < \cdots < x_m \leqslant 1$$

It is clear that such a formula is determined uniquely by the coefficients $\{A_i\}_{i=1}^{m}$ and the numbers $\{\gamma_i\}_{i=1}^{m}$, $\gamma_i = x_i$, $i = 1, 2, \ldots, m$, which give the position of the knots $\{x_i\}_1^m$ in $[0, 1]$.

We shall call the collection of numbers $\{\{A_i\}_1^m, \{\gamma_i\}_1^m\} = \{A, \gamma\}$ the type of the quadrature formula Lf. Therefore the type is determined by the number m and the collections $\{A_i\}_1^m$, $\{\gamma_i\}_1^m$. From a quadrature formula Lf in the interval $[0, 1]$ of a given type $\{A, \gamma\}$ we can obtain a quadrature formula for an arbitrary finite interval $[a, b]$ in the following way: we multiply the coefficients A_i by $(b - a)$, i.e. they become $(b - a)A_i$, $i = 1, 2, \ldots, m$, and the knots x_i, $i = 1, 2, \ldots, m$, are determined by $x_i = a + \gamma_i(b - a)$, $i = 1, 2, \ldots, m$. So for an arbitrary type $\{A, \gamma\}$, where the numbers γ satisfy the inequalities $0 \leqslant \gamma_1 < \gamma_2 < \cdots < \gamma_m \leqslant 1$, and for every finite interval $[a, b]$, we have a quadrature formula $L(\{A, \gamma\}, [a, b])f$ of type $\{A, \gamma\}$ which is given by

$$L(\{A, \gamma\}, [a, b])f = \sum_{i=1}^{n} (b - a)A_i f(x_i) \quad x_i = a + \gamma_i(b - a) \quad (3.20)$$

Obviously for every interval $[a,b]$ we have

$$\|L(\{A,\gamma\},[a,b])\|_{M\to M} = (b-a)\sum_{i=1}^{n}|A_i| = (b-a)\|L(\{A,\gamma\},[0,1])\|_{M\to M} \quad (3.21)$$

Definition 3.1. Let Lf be a quadrature formula of type $\{A,\gamma\}$. We shall say that the quadrature formula $\mathscr{L}_n(f)$ is an n-composite quadrature formula in the interval $[a,b]$, generated by the quadrature formula Lf if we have

$$\mathscr{L}_n(f) = \sum_{i=1}^{n} L(\{A,\gamma\}, [x_{i-1}, x_i])f \quad (3.22)$$

where $x_i = a + i(b-a)/n$, $i = 0, 1, \ldots, n$.

So the n-composite quadrature formula $\mathscr{L}_n(f)$, generated by the quadrature formula L of type $\{A,\gamma\}$ can be obtained in the following way: we divide the interval $[a,b]$ into n equal intervals $[x_{i-1}, x_i]$, $i = 1, 2, \ldots, n$, and for each of the intervals we apply to f the quadrature formula of type $\{A,\gamma\}$ (which is uniquely determined by (3.20) according to the type and the interval).

Remark. Of course, it is possible to consider the more general formula

$$\mathscr{L}_n(f) = \sum_{i=1}^{n} L(\{A^{(i)}, \gamma^{(i)}\}, [x_{i-1}, x_i])f$$

where $\{A^{(i)}, \gamma^{(i)}\}$, $i = 1, 2, \ldots, n$, are different types of quadrature formulae and $\{x_i\}_0^n$ is an arbitrary partition of the interval $[a,b]$ (where, for the different types of quadrature, m can also be different). The results which we shall obtain for the n-composite quadrature formulae can be generalized in an obvious way for such formulae. Since the formulation of such general results is very complex, and since usually in practice we use only composite quadratures of the type of equation (3.22), we shall restrict ourselves only to estimations of the quadrature formulae of the type of equation (3.22).

Definition 3.2. We shall say that the quadrature formula Lf in the interval $[a,b]$ has precision k, if for every polynomial $f \in H_k$ we have

$$\int_a^b f(x)\,dx = Lf \quad (3.23)$$

and there exists an algebraic polynomial $g \in H_{k+1}$ for which (3.23) fails.

Similarly, if we consider quadrature formulae for numerical evaluation of integrals with weight p, i.e.

$$\int_a^b p(x)f(x)\,dx$$

we shall say that the corresponding formula Lf has precision k, if for every $f \in H_k$

we have

$$\int_a^b p(x)f(x)\,dx = Lf \tag{3.24}$$

and there exists an algebraic polynomial $f \in H_{k+1}$ for which (3.24) is not true.

Theorem 3.4. *Let $\mathscr{L}_n(f)$ be an n-composite quadrature formula in the interval $[a,b]$, generated by the quadrature formula Lf of type $\{A,\gamma\}$ and with precision k. Then*

$$\left| \int_a^b f(x)\,dx - \mathscr{L}_n(f) \right| \leqslant (b-a)W_{k+1}\left(1 + \sum_{i=1}^m |A_i|\right)\tau_{k+1}\left(f; \frac{2(b-a)}{n(k+1)}\right)_L \tag{3.25}$$

If Lf, has precision k with respect to the weight p, then

$$\left| \int_a^b p(x)f(x)\,dx - Lf \right| \leqslant CW_{k+1}(f;b-a)_C \tag{3.26}$$

Remark. The estimates (3.25) and (3.26) do not depend explicitly on γ, but their dependence on γ is through the precision of Lf.

Proof of Theorem 3.4. Let $x_i = a + i(b-a)/n$, $i = 0, 1, \ldots, n$. From (3.22) we have

$$\left| \int_a^b f(x)\,dx - \mathscr{L}_n(f) \right| = \left| \sum_{i=1}^n \left[\int_{x_{i-1}}^{x_i} f(x)\,dx - L(\{A,\gamma\}, [x_{i-1}, x_i])f \right] \right| \tag{3.27}$$

Since the quadrature formula $L(\{A,\gamma\}, [x_{i-1}, x_i])$ has precision k in the interval $[x_{i-1}, x_i]$, then for every $q \in H_k$ we have

$$\int_{x_{i-1}}^{x_i} q(x)\,dx - L(\{A,\gamma\}, [x_{i-1}, x_i])q = 0 \tag{3.28}$$

Let us consider the linear functionals $S_i f$:

$$S_i f = \int_{x_{i-1}}^{x_i} f(x)\,dx - L(\{A,\gamma\}, [x_{i-1}, x_i])f \quad i = 1, 2, \ldots, n$$

The functional S_i is defined for all bounded and integrable (measurable) functions in the interval $[x_{i-1}, x_i]$ and vanishes for all algebraic polynomials of degree k due to (3.28). If we apply Theorem 2.4 to the functionals S_i, we obtain

$$|S_i f| \leqslant \|S_i\|_M W_{k+1}\omega_{k+1}(f; [x_{i-1}, x_i]) \tag{3.29}$$

For the modulus of continuity of the function f in the interval $[x_{i-1}, x_i]$ we have

$$\omega_{k+1}(f; [x_{i-1}, x_i]) = \omega_{k+1}\left(f, \frac{x_{i-1} + x_i}{2}; \frac{b-a}{n(k+1)}\right) \tag{3.30}$$

For $\|S_i\|_M$ we obtain, using (3.21),

$$|S_i f| \leqslant (x_i - x_{i-1}) \| f \|_{C[x_{i-1}, x_i]} + (x_i - x_{i-1}) \sum_{i=1}^{m} |A_i| \| f \|_{C[x_{i-1}, x_i]} \quad (3.31)$$

i.e.

$$\|S_i\|_M \leqslant (x_i - x_{i-1}) \left(1 + \sum_{i=1}^{m} |A_i| \right) \quad (3.32)$$

From (3.27) and (3.29)–(3.32) we obtain

$$\left| \int_a^b f(x) \, dx - \mathscr{L}_n(f) \right| \leqslant \sum_{i=1}^{n} \left| \int_{x_{i-1}}^{x_i} f(x) \, dx - L(\{A, \gamma\}, [x_{i-1}, x_i]) f \right|$$

$$\leqslant \sum_{i=1}^{n} (x_i - x_{i-1}) W_{k+1} \left(1 + \sum_{j=1}^{m} |A_j| \right) \omega_{k+1}\left(f, \frac{x_{i-1} + x_i}{2}; \frac{b-a}{n(k+1)} \right)$$

$$= W_{k+1} \left(1 + \sum_{j=1}^{m} |A_j| \right) \sum_{i=1}^{n} \int_{x_{i-1}}^{x_i} \omega_{k+1}\left(f, \frac{x_{i-1} + x_i}{2}; \frac{b-a}{n(k+1)} \right) dx$$

$$\leqslant W_{k+1} \left(1 + \sum_{j=1}^{m} |A_j| \right) \sum_{i=1}^{n} \int_{x_{i-1}}^{x_i} \omega_{k+1}\left(f, x; \frac{2(b-a)}{n(k+1)} \right) dx \quad (3.33)$$

since for every $x \in [x_{i-1}, x_i]$ we have

$$\omega_{k+1}\left(f, \frac{x_{i-1} + x_i}{2}; \frac{b-a}{n(k+1)} \right) \leqslant \omega_{k+1}\left(f, x; \frac{2(b-a)}{n(k+1)} \right).$$

The inequality (3.33) then gives

$$\left| \int_a^b f(x) \, dx - \mathscr{L}_n(f) \right| \leqslant W_{k+1} \left(1 + \sum_{j=1}^{m} |A_j| \right) \int_a^b \omega_{k+1}\left(f, x; \frac{2(b-a)}{n(k+1)} \right) dx$$

$$= (b-a) W_{k+1} \left(1 + \sum_{j=1}^{m} |A_j| \right) \tau_{k+1}\left(f; \frac{2(b-a)}{n(k+1)} \right)_L.$$

Inequality (3.26) follows from Theorem 2.4.   ∎

Remark. The estimate (3.25) permits us, using the properties of the averaged moduli $\tau_{k+1}(f; \delta)_L$, to obtain several corollaries concerning the rate of convergence of the quadrature process $\mathscr{L}_n(f)$ as $n \to \infty$. However, as a consequence of its generality, this estimate in the general case is worse, up to a multiplicative constant, than the best possible estimate for a concrete composite quadrature formula (compare with the previous section). Let us note two things which make the constant larger: (a) the estimate of $\|S_i\|_M$; (b) replacing $\omega_k(f; [x_{i-1}, x_i])$ by $\omega_k(f, x; (b-a)2/n(k+1))$. The first is more important. It depends directly on Whitney's constant W_{k+1}.

Let $\mathscr{L}_n(f)$ be an n-composite quadrature formula for the interval $[a, b]$, generated by a quadrature formula Lf with precision k. Then for the error

$$R_n(f) = \left| \int_a^b f(x) \, dx - \mathscr{L}_n(f) \right|$$

using Theorem 3.4 and the properties of the averaged moduli, we have the asymptotic estimates given in the table below.

f	$R_n(f)$
Riemann integrable	$o(1)$
$f^{(i)}$, $i = 0, 1, \ldots, k$, has bounded variation	$O(n^{-i-1})$
$f^{(i)}$, $i = 0, 1, \ldots, k$ is absolutely continuous	$o(n^{-i-1})$

3.2.2. Estimates for some classical quadrature formulae

We shall begin with the composite Newton–Cotes formulae. These are obtained by dividing the interval $[a, b]$ into equal subintervals by means of the points $x_i = a + ih$, $i = 0, 1, \ldots, n$, $h = (b - a)/n$. For every interval $[x_{i-1}, x_i]$ the integral

$$\int_{x_{i-1}}^{x_i} f(x)\,dx$$

is replaced by

$$\int_{x_{i-1}}^{x_i} P_{i,k}(x)\,dx,$$

where $P_{i,k}$ is the interpolation polynomial for the function f of kth degree with knots at the points

$$y_{i,j} = x_{i-1} + j\frac{x_i - x_{i-1}}{k} \quad j = 0, 1, \ldots, k$$

for the Newton–Cotes formulae of closed type; or with knots at the points

$$y_{i,j} = x_{i-1} + \frac{j(x_i - x_{i-1})}{k+2} \quad j = 1, 2, \ldots, k+1$$

for the Newton–Cotes formulae of open type. It is well known (see Berezin and Zhidkov [1], Sendov and Popov [1]) that the quadrature

$$\int_{x_{i-1}}^{x_i} f(x)\,dx = \int_{x_{i-1}}^{x_i} P_{i,k}(x)\,dx + R_{i,k}(f)$$

has precision k for odd k and precision $k + 1$ for even k.
   The composite Newton–Cotes formula has the form

$$\int_a^b f(x)\,dx = \sum_{i=1}^n \int_{x_{i-1}}^{x_i} P_{i,k}(x)\,dx + R_n^k(f)$$

$$= \frac{b-a}{n} \sum_{i=1}^n \sum_{j=\varepsilon}^{k+\varepsilon} A_{k,j} f(y_{i,j}) + R_n^k(f)$$

where $\varepsilon = 0$ for the formulae of closed type and $\varepsilon = 1$ for the formulas of open type.

Using Theorem 3.4 we obtain the following.

Theorem 3.5. *For the error of the Newton–Cotes quadrature formula we have the estimates:*
(a) *for odd* k:

$$|R_n^k(f)| \leqslant \left(1 + \sum_{j=\varepsilon}^{k+\varepsilon} |A_{k,j}|\right) W_{k+1}\tau_{k+1}\left(f; \frac{2(b-a)}{n(k+1)}\right)_L$$

(b) *for even* k:

$$|R_n^k(f)| \leqslant \left(1 + \sum_{j=\varepsilon}^{k+\varepsilon} |A_{k,j}|\right) W_{k+2}\tau_{k+2}\left(f; \frac{2(b-a)}{n(k+2)}\right)_L$$

We mention that these estimates have corollaries which are similar to the corollaries of the general Theorem 3.4. Without formulating these consequences exactly, we note that for functions having derivatives of order $< k + 1$ (respectively $< k + 2$) we obtain new results. For functions which have bounded $(k + 1)$th (respectively $(k + 2)$th) derivative, the estimates of Theorem 3.5 give the same order as the classical estimates up to a multiplicative constant. If we want to obtain better constants (or the exact constant) we must consider the concrete quadrature formulae.

In an analogous way we can investigate all composite quadrature formulae. We shall restrict ourselves only to the formulation of the result for the composite quadrature formulae of Gaussian type with weight $p \equiv 1$ (see Sendov and Popov [1], Bahvalov [1]).

Theorem 3.6. *Let* $\mathscr{L}_n(f)$ *be an* n-*composite quadrature formula in the interval* $[a, b]$, *generated by a Gaussian quadrature formula of* kth *order (i.e. a Gaussian quadrature formula with* k *knots) with weight* $p \equiv 1$. *Then*

$$\left|\int_a^b f(x)\,dx - \mathscr{L}_n(f)\right| \leqslant (b-a) W_{2k}\tau_{2k}\left(f; \frac{(b-a)}{nk}\right)_L$$

This estimate can be obtained using the general Theorem 3.4. It is a well-known fact that the Gaussian quadrature formula with k knots has precision $2k - 1$, and also that for weight $p \equiv 1$ all coefficients in the Gaussian formula are positive and their sum is 1.

We shall now show that the estimate of Theorem 3.1

$$|R_n^0(f)| \leqslant \frac{1}{2}\tau_2\left(f; \frac{b-a}{2n}\right)_L \tag{3.34}$$

is exact for the class of all infinitely differentiable functions, and therefore it is exact for the class of all bounded measurable functions (Ivanov [1]).

Let us consider the sequences of functions $\{f_m\}_1^\infty$ in the interval $[0, 1]$, defined by

$$f_m(x) = \varphi\left[2mn\left(x - \frac{2i-1}{2n}\right)\right] \quad \left|x - \frac{2i-1}{2n}\right| \leqslant \frac{1}{2n}, \quad i = 1, 2, \ldots, n$$

where

$$\varphi(x) = \begin{cases} 0 & |x| > 1 \\ \exp\left(\dfrac{-x^2}{1-x^2}\right) & |x| \leqslant 1 \end{cases}$$

For the functions f_m, $m = 1, 2, \ldots$, we have

$$f_m \in C^{\infty}_{[0,1]} \quad 0 \leqslant \int_0^1 f_m(x)\,dx \leqslant \frac{1}{m}$$

$$\frac{1}{n}\sum_{i=1}^{n} f_m\left(\frac{2i-1}{2n}\right) = 1$$

and therefore

$$|R_n^0(f_m)| \geqslant 1 - \frac{1}{m}$$

For every m we have $\tau_2(f_m; 1/2n)_L \leqslant 2$ since $\omega_2(f_m, x; 1/2n) \leqslant 2$ for every $x \in [0, 1]$. Therefore

$$\frac{|R_n^0(f_m)|}{\tau_2(f_m; 1/2n)_L} \geqslant \frac{1 - 1/m}{2}$$

from which we obtain

$$\sup_{f \in C^{\infty}_{[0,1]}} \frac{|R_n^0(f)|}{\tau_2(f; 1/2n)_L} \geqslant \tfrac{1}{2}$$

This inequality shows that the constant $\tfrac{1}{2}$ before τ_2 in (3.34) is exact.

To show that the constant $(b-a)/2n$ in τ_2, in the estimate (3.34), is exact also, it is sufficient to note that, for every number $c_1 < \tfrac{1}{2}$, we have

$$\tau_2\left(f_m; \frac{c_1}{n}\right)_L \leqslant 2 - \frac{\tfrac{1}{2} - c_1}{4} \quad \text{for sufficiently large } m$$

and therefore the assumption that

$$|R_n^0(f_m)| \leqslant \tfrac{1}{2}\tau_2\left(f_m; \frac{c_1}{n}\right)_L$$

leads us to the contradiction

$$1 - 1/m \leqslant 1 - \frac{\tfrac{1}{2} - c_1}{8}$$

for sufficiently large m.

With more precise computation we can improve the constant for the composite trapezoidal rule. We have

$$\int_{x_{i-1}}^{x_i} \left[-f(x) + \frac{f(x_{i-1}) + f(x_i)}{2} \right] dx = \frac{f(x_{i-1})b - a}{2n} + \frac{1}{2} \int_{x_{i-1}}^{x_i} f(x) dx$$

$$- 2 \int_{x_{i-1}}^{(x_{i-1} + x_i)/2} f(x) dx + \frac{f(x_i)(b - a)}{2n} + \frac{1}{2} \int_{x_{i-1}}^{x_i} f(x) dx - 2 \int_{(x_{i-1} + x_i)/2}^{x_i} f(x) dx$$

$$= \int_0^{h/2} [f(x_{i-1}) - 2f(x_{i-1} + t) + f(x_{i-1} + 2t)] dt$$

$$+ \int_0^{h/2} [f(x_i) - 2f(x_i - t) + f(x_i - 2t)] dt$$

Thus

$$\left| \int_{x_{i-1}}^{x_i} \left[f(x) - \frac{f(x_{i-1}) + f(x_i)}{2} \right] dx \right| \leq \int_0^{h/2} \omega_2 \left(f, x_{i-1} + t; \frac{h}{2} \right) dt$$

$$+ \int_0^{h/2} \omega_2 \left(f, x_i - t; \frac{h}{2} \right) dt$$

$$= \int_{x_{i-1}}^{x_i} \omega_2 \left(f, x; \frac{h}{2} \right) dx$$

Therefore (compare with Theorem 3.2)

$$|R_n^1(f)| \leq \tau_2 \left(f; \frac{b - a}{2n} \right)_L$$

3.3. An estimate of the error of composite quadrature formulae, using values of the integrand and its derivatives

3.3.1. General estimation

To obtain an estimate of the error of composite quadrature formulae which use values of the integrand and its derivatives, we need a generalization of Theorem 2.4. We shall restrict ourselves to one lemma, which will be sufficient for our purpose.

Let Lf be a quadrature formula in $[a, b]$, using the values of the function f at the points y_i, $i = 0, 1, \ldots, m$, and the values of the derivatives of f up to the order α_i at the points y_i, $i = 0, 1, \ldots, m$. Such formulae usually have the form

$$Lf = \sum_{i=0}^{m} \sum_{j=0}^{\alpha_i} A_{ij}(b - a)^{j+1} f^{(j)}(y_i) \tag{3.35}$$

i.e. they are obtained from the corresponding quadrature formula for the interval $[0, 1]$ on multiplying the coefficients by $(b - a)^s$ and making the corresponding change of the knots.

Let $R(f)$ be the error of such a quadrature formula:

$$R(f) = \int_a^b f(x)\,dx - Lf$$

We shall prove the following lemma.

Lemma 3.1. *Let the quadrature formula Lf have precision k, i.e. $R(f)=0$ if $f \in H_k$. Then for every function f, which has bounded derivatives up to the order $r = \max\{\alpha_i: 0 \leqslant i \leqslant m\}$, $k \geqslant r$, the following estimate holds:*

$$|R(f)| \leqslant (b-a)^{r+1} W_{k+1-r} \left(\frac{1}{r!} + \sum_{i=0}^m \sum_{j=0}^{\alpha_i} \frac{|A_{ij}|}{(r-j)!} \right) \omega_{k+1-r}(f^{(r)};[a,b])$$

Proof. Let the function f have bounded derivatives up to the order $r = \max\{\alpha_i: 0 \leqslant i \leqslant m\}$. From Theorem 2.2, for $k \geqslant r$ and $P = T_{k-r}(f^{(r)}) \in H_{k-r}$ (see Notation 11) we obtain

$$\|P - f^{(r)}\|_{M[a,b]} \leqslant W_{k+1-r}\omega_{k+1-r}(f^{(r)};[a,b]) \tag{3.36}$$

Let us consider the polynomial $Q \in H_k$*

$$Q(x) = f(a) + \frac{x-a}{1!}f'(a) + \cdots + \frac{(x-a)^{r-1}}{(r-1)!}f^{(r-1)}(a)$$

$$+ \frac{1}{(r-1)!}\int_a^x (x-t)^{r-1}P(t)\,dt \tag{3.37}$$

and the identity

$$f(x) = f(a) + \frac{x-a}{1!}f'(a) + \cdots + \frac{(x-a)^{r-1}}{(r-1)!}f^{(r-1)}(a)$$

$$+ \frac{1}{(r-1)!}\int_a^x (x-t)^{r-1}f^{(r)}(t)\,dt \tag{3.38}$$

From (3.36)–(3.38) we obtain immediately that

$$|f^{(j)}(y_i) - Q^{(j)}(y_i)| \leqslant \frac{(b-a)^{r-j}}{(r-j)!}W_{k+1-r}\omega_{k+1-r}(f^{(r)};[a,b]) \tag{3.39}$$

where

$$i = 0, 1, \ldots, m, \quad j = 0, 1, \ldots, \alpha_i.$$

Since $Q \in H_k$, we have $R(Q) = 0$ and hence from (3.39)

$$|R(f)| \leqslant |R(f-Q)| + |R(Q)|$$
$$= |R(f-Q)|$$

---

*$Q \in H_k$, since $Q^{(r)} = P \in H_{k-r}$.

$$\leqslant \int_a^b |f(x) - Q(x)| dx + \sum_{i=0}^m \sum_{j=0}^{\alpha_i} |A_{ij}|(b-a)^{j+1}|f^{(j)}(y_i) - Q^{(j)}(y_i)|$$

$$\leqslant \int_a^b |f(x) - Q(x)| dx + (b-a)^{r+1} W_{k+1-r} \omega_{k+1-r}(f^{(r)}; [a,b]) \sum_{i=0}^m \sum_{j=0}^{\alpha_i} \frac{|A_{ij}|}{(r-j)!}$$

(3.40)

From (3.36)–(3.38) it follows that for every $x \in [a,b]$

$$|f(x) - Q(x)| \leqslant \frac{(b-a)^r}{r!} W_{k+1-r} \omega_{k+1-r}(f^{(r)}; [a,b])$$

which, together with (3.40), gives

$$|R(f)| \leqslant (b-a)^{r+1} W_{k+1-r} \left[\frac{1}{r!} + \sum_{i=0}^m \sum_{j=0}^{\alpha_i} \frac{|A_{ij}|}{(r-j)!} \right] \omega_{k+1-r}(f^{(r)}; [a,b]) \quad \blacksquare$$

Using Lemma 3.1 we can easily obtain estimations for the error of composite quadrature formulae, generated by quadratures of the type (3.35), using the averaged moduli of smoothness. More precisely let

$$Lf = \sum_{i=0}^m \sum_{j=0}^{\alpha_i} A_{ij} f^{(j)}(x_i) \quad 0 \leqslant x_0 < x_1 < \cdots < x_m \leqslant 1 \tag{3.41}$$

be a quadrature formula in the interval $[0,1]$, which uses the values of the function f at the points x_i and the values of the derivatives of f at the points x_i up to the order α_i, $r = \max\{\alpha_i : 0 \leqslant i \leqslant m\}$. From (3.41) we obtain a quadrature formula of the same type for the arbitrary interval $[c,d]$ using

$$L(f; [c,d]) = \sum_{i=0}^m \sum_{j=0}^{\alpha_i} A_{ij}(d-c)^{j+1} f^{(j)}(y_i), \tag{3.42}$$

where $y_i = c + x_i(d-c)$, $i = 0, 1, \ldots, m$. Let us note that if Lf has precision k, then for the arbitrary interval $[c,d]$ the quadrature formula (3.42) also has the same precision.

As in the previous section, we say that $\mathscr{L}_n(f)$ is an n-composite quadrature for the interval $[a,b]$, generated by the quadrature (3.41) if

$$\mathscr{L}_n(f) = \sum_{i=1}^n L(f; [x_{i-1}, x_i])$$

where $x_i = a + i(b-a)/n$, $i = 0, 1, \ldots, n$, and $L(f; [x_{i-1}, x_i])$ are given by (3.42).

Theorem 3.7. *Let $\mathscr{L}_n(f)$ be an n-composite quadrature formula for the interval $[a,b]$, generated by the quadrature formula (3.41). Let (3.41) have precision k, $k \geqslant r$. Then the following estimate holds:*

$$|R_n(f)| = \left| \int_a^b f(x) dx - \mathscr{L}_n(f) \right|$$

$$\leqslant W_{k+1-r} \left(\frac{b-a}{n} \right)^r \left[\frac{1}{r!} + \sum_{i=0}^m \sum_{j=0}^{\alpha_i} \frac{|A_{ij}|}{(r-j)!} \right] \tau_{k+1-r} \left(f^{(r)}; \frac{2(b-a)}{n(k+1-r)} \right)_L$$

Proof. Using the definition of the n-composite quadrature formula, generated by the quadrature formula Lf, we obtain

$$|R_n(f)| \leq \sum_{i=1}^{n} \left| \int_{x_{i-1}}^{x_i} f(x)\,dx - L(f;[x_{i-1},x_i]) \right| \qquad (3.43)$$

Since $L(f;[x_{i-1}, x_i])$ has precision k and has the form (3.35), we can apply Lemma 3.1. Since $x_i - x_{i-1} = (b-a)/n$, we obtain

$$\left| \int_{x_{i-1}}^{x_i} f(x)\,dx - L(f;[x_{i-1},x_i]) \right|$$

$$\leq \left(\frac{b-a}{n}\right)^{r+1} W_{k+1-r}\left(\frac{1}{r!} + \sum_{i=0}^{m}\sum_{j=0}^{\alpha_i}\frac{|A_{ij}|}{(r-j)!}\right)\omega_{k+1-r}(f^{(r)};[x_{i-1},x_i]) \qquad (3.44)$$

If $x \in [x_{i-1}, x_i]$, then

$$\omega_{k+1-r}(f^{(r)};[x_{i-1},x_i]) \leq \omega_{k+1-r}\left(f^{(r)},x;\frac{2(b-a)}{n(k+1-r)}\right)$$

This inequality, together with (3.44) gives

$$\left| \int_{x_{i-1}}^{x_i} f(x)\,dx - L(f;[x_{i-1},x_i]) \right|$$

$$\leq \left(\frac{b-a}{n}\right)^{r} W_{k+1-r}\left(\frac{1}{r!} + \sum_{i=0}^{m}\sum_{j=0}^{\alpha_i}\frac{|A_{ij}|}{(r-j)!}\right)\int_{x_{i-1}}^{x_i}\omega_{k+1-r}\left(f^{(r)},x;\frac{2(b-a)}{n(k+1-r)}\right)dx$$

$$(3.45)$$

The statement of the theorem follows from (3.43)–(3.45):

$$|R_n(f)| \leq \left(\frac{b-a}{n}\right)^{r} W_{k+1-r}\left(\frac{1}{r!} + \sum_{i=0}^{m}\sum_{j=0}^{\alpha_i}\frac{|A_{ij}|}{(r-j)!}\right)\int_{a}^{b}\omega_{k+1-r}\left(f^{(r)},x;\frac{2(b-a)}{n(k+1-r)}\right)dx$$

$$= W_{k+1-r}\frac{(b-a)^r}{n^r}\left(\frac{1}{r!} + \sum_{i=0}^{m}\sum_{j=0}^{\alpha_i}\frac{|A_{ij}|}{(r-j)!}\right)\tau_{k+1-r}\left(f^{(r)};\frac{2(b-a)}{n(k+1-r)}\right)_L$$

∎

From Theorem 3.7, using the properties of the averaged moduli of smoothness, we obtain corollaries given in the table below.

$f^{(r)}$	$R_n(f)$
Riemann integrable	$o(n^{-r})$
bounded variation	$O(n^{-r-1})$
absolutely continuous	$o[n^{-r-1}\omega_{k-r}(f^{(r+1)};n^{-1})_L]$

Remark. The estimate of Theorem 3.7, which we can write in the form

$$|R_n(f)| = O[n^{-r}\tau_{k+1-r}(f^{(r)}; n^{-1})_L]$$

cannot be improved in the sense that we cannot replace $\tau_{k+1-r}(f^{(r)}; n^{-1})_L$ by the expression $n\tau_{k+2-r}(f^{(r-1)}; n^{-1})_L$. This follows from, the fact that in the local estimate (see Lemma 3.1), we cannot replace $(b-a)^{-1}\omega_{k+2-r}(f^{(r-1)}; [a,b])$ by $\omega_{k+1-r}(f^{(r)}; [a,b])$. We shall show this by the following example due to Ivanov.

Let us consider the quadrature formula

$$Lf = \tfrac{1}{6}[4f(0) + 2f(1) + f'(0)]$$

for the interval $[0,1]$. It is easy to see that Lf has precision 2.
From Lemma 3.1 we have, for the error

$$R(f) = \int_0^1 f(x)\,dx - Lf$$

the estimate:

$$|R(f)| = O[\omega_2(f'; [0,1])]$$

We shall show that it is not possible to obtain an estimate of the type

$$|R(f)| = O[\omega_3(f; [0,1])] \tag{3.46}$$

Let us consider the following sequences of functions:

$$f_n(x) = \begin{cases} 0 & \text{if } 0 \leqslant x \leqslant \dfrac{1}{n} - \dfrac{1}{4(n-1)} \\ n\left(x - \dfrac{1}{n} + \dfrac{1}{4(n-1)}\right)^2 & \text{if } \left|x - \dfrac{1}{n}\right| \leqslant \dfrac{1}{4(n-1)} \\ \dfrac{nx-1}{n-1} & \text{if } \dfrac{1}{n} + \dfrac{1}{4(n-1)} \leqslant x \leqslant 1 \end{cases}$$

for $n = 2, 3, \ldots$.

Obviously $f_n \in C^1[0,1]$. For every $x \in [0,1]$ we have $(nx-1)/(n-1) \leqslant f_n(x) \leqslant x$ and $f_n(x) \geqslant 0$. Therefore, if $h > 0$, $x, x + 3h \in [0,1]$, then

$$\Delta_h^3 f_n(x) = f_n(x+3h) - 3f_n(x+2h) + 3f_n(x+h) - f_n(x)$$

$$\leqslant x + 3h - 3\frac{n(x+2h)-1}{n-1} + 3(x+h) - \frac{nx-1}{n-1} \leqslant \frac{4}{n-1}$$

$$\Delta_h^3 f_n(x) \geqslant -x + 3\frac{n(x+h)-1}{n-1} - 3(x+2h) + \frac{n(x+3h)-1}{n-1} \geqslant -\frac{4}{(n-1)}$$

i.e.

$$\omega_3(f_n; [0,1]) \leqslant \frac{4}{n-1}$$

However
$$Lf_n = \tfrac{1}{6}(4f(0) + 2f(1) + 1f'(0)) = \tfrac{1}{3}$$

and
$$\int_0^1 f_n(x)\,dx = \frac{n-1}{2n} + \frac{1}{96}(n-1)^3 > \frac{n-1}{2n}$$

Therefore
$$\frac{|\int_0^1 f_n(x)\,dx - Lf_n|}{\omega_3(f_n;[0,1])} > \frac{(n-1)(n-3)}{24n}$$

which contradicts the estimate (3.46).

3.3.2. The Obreshkov–Chakalov quadrature formulae

The Obreshkov–Chakalov quadrature formulae use the values of the function and its consecutive derivatives at both ends of the interval over which we integrate. For the interval $[0,1]$ they have the form

$$L_{l,m}(f) = \sum_{j=0}^{l} A_j f^{(j)}(0) + \sum_{j=0}^{m} B_j f^{(j)}(1)$$

where the coefficients A_j, B_j are given by

$$A_j = \frac{(-1)^{l+m}}{j!} \sum_{i=0}^{l-j} \binom{m+i}{i} \frac{(l+1)!(j+1)!}{(l+j+i+2)!}$$

$$B_j = \frac{(-1)^j}{j!} \sum_{i=0}^{m-j} \binom{l+i}{i} \frac{(m+1)!(j+1)!}{(m+j+i+2)!}$$

We shall call this the Obreshkov–Chakalov quadrature formula of the type (l, m).

In the particular case when $m = l = r$, we have

$$A_j = \frac{1}{(j+1)!} \binom{r+1}{j+1} \Big/ \binom{2r+2}{j+1}$$

$$B_j = (-1)^j A_j$$

Since the Obreshkov–Chakalov quadrature formulae have precision $l+m+1$, we obtain the following theorem from Theorem 3.7.

Theorem 3.8. *Let* $\mathscr{L}_n(f)$ *be the n-composite quadrature formula, generated by the Obreshkov–Chakalov quadrature formula of the type* (l, m). *Then for the error*

$$R_n(f) = \int_a^b f(x)\,dx - \mathscr{L}_n(f)$$

we have the estimate

$$|R_n(f)| \leqslant W_{l+m+2-r} \frac{(b-a)^r}{n^r} \tau_{l+m+2-r}\left(f^{(r)}; \frac{b-a}{n}\right)_L,$$

where $r = \max\{l, m\}$.

   In the particular case when $l = m = r$, *we have*

$$|R_n(f)| \leqslant W_{r+2} \frac{(b-a)^r}{n^r} \tau_{r+2}\left(f^{(r)}; \frac{b-a}{n}\right)_L.$$

3.3.3.   The Hermite formulae

The Hermite quadrature formulae are obtained by replacing the integrand by its Hermite interpolation polynomial. The values of the Hermite interpolation polynomial coincide with the values of the function and its derivative at $k+1$ equidistant points. For the interval $[0, 1]$ they have the form

$$L_k(f) = \sum_{j=0}^{k}\left[A_j f\left(\frac{j}{k}\right) + B_j f'\left(\frac{j}{k}\right)\right] \tag{3.47}$$

where

$$A_j = \int_0^1 \left[1 - \frac{\varphi''(j/k)(x-j/k)}{\varphi'(j/k)}\right]\left[\frac{\varphi(x)}{(x-j/k)\varphi'(j/k)}\right]^2 dx$$

$$B_j = \int_0^1 \frac{\varphi^2(x)}{(x-j/k)[\varphi'(j/k)]^2} dx \quad \varphi(x) = x(x-1/k)(x-2/k)\cdots(x-1)$$

   The quadrature formula (3.47) has precision $2k+1$. Using Theorem 3.7, we obtain the following result.

Theorem 3.9.   *Let* $\mathcal{L}_n(f)$ *be an n-composite quadrature formula for the interval* $[a, b]$, *which is generated by the Hermite quadrature formula* (3.47). *Then for the error*

$$R_n(f) = \int_a^b f(x)\,dx - \mathcal{L}_n(f)$$

we have

$$|R_n(f)| \leqslant W_{2k+1}(b-a)n^{-1}\tau_{2k+1}\left(f'; \frac{2(b-a)}{(2k+1)n}\right)_L.$$

Let us emphasize once more that the problem of obtaining the exact values of the constants in each theorem of this section remains open.

3.4.   An estimate of the error of quadrature formulae for periodic functions

In the previous sections we have obtained estimates of the error of nth composite quadrature formulae. If the generated quadrature formula has precision k and the function f has rth derivative of bounded variation, then for the error $R_n(f)$ of the nth composite quadrature formula we have the estimate

$$R_n(f) = O(n^{-\min(k+1,r+1)})$$

We see that the precision of the generated quadrature formula plays an important role in the estimate of the error. For periodic functions this is not so. If the generated quadrature formula is exact only for the constant function then the error has order $O(n^{-r-1})$ if f has rth derivative of bounded variation. We shall give this result here, which is due to Ivanov [2].

Let $\mathscr{L}_n(f)$ be an n-composite quadrature formula for the interval $[0, 2\pi]$, generated by a quadrature formula Lf of the type

$$\{A, \alpha\} = \{\{A_j\}_{j=1}^m, \{\alpha_j\}_{j=1}^m\}$$

which is exact for the constant function, i.e. we have

$$\mathscr{L}_n(f) = \sum_{i=1}^n L(f; [x_{i-1}, x_i])$$

$$= \sum_{i=1}^n \sum_{j=1}^m \frac{2\pi}{n} A_j f[x_{i-1} + \alpha_j(x_i - x_{i-1})] \tag{3.48}$$

where $x_i = 2\pi i/n, i = 0, 1, \ldots, n, \alpha_j \in [0, 1], j = 1, 2, \ldots, m,$

$$L(\{A, \alpha\}; [0, 2\pi])C = 2\pi \sum_{j=1}^m A_j C = 2\pi C$$

We shall obtain an estimate for the error

$$R_n(f) = \int_0^{2\pi} f(x)\,dx - \mathscr{L}_n(f)$$

Let $\tilde{W}^r L(M; 0, 2\pi)$ denote the set of all 2π-periodic functions, for which $f^{(r-1)}$ is absolutely continuous and $\|f^{(r)}\|_{L[0,2\pi]} \leq M, r = 1, 2, \ldots$.
Let B be a class of continuous 2π-periodic functions with the properties:

(1) If $f_1, f_2 \in B$ then $\alpha f_1 + (1 - \alpha)f_2 \in B$ for $\alpha \in [0, 1]$.
(2) If $f \in B$ then $-f, f + C, f(x + a) \in B$ for every constant C and a.

For example the classes $\tilde{W}^r L(M; 0, 2\pi), r = 1, 2, \ldots$, are such classes of functions.
For every natural number n let us denote

$$B_m = \{f \in B: f \text{ is } (2\pi/n)\text{-periodic and } \int_0^{2\pi} f(x)dx = 0\}$$

$$\mu_n[B] = \sup\{\|f\|_{C[0,2\pi]}: f \in B_n\}$$

$$= \sup\{|f(0)|: f \in B_n\}$$

$$R_n[\mathscr{D}] = \sup\{|R_n(f)|: f \in \mathscr{D}\}$$

We shall prove the following lemma, which is analogous to a lemma of Motornii (see Nikolskii [1], p. 208, Lemma 18).

Lemma 3.2. *We have*

$$R_n[B] \leqslant 2\pi\mu_n[B] \sum_{j=1}^{m} |A_j|$$

Proof. Since Lf is exact for the constant function $\mathscr{L}_n(f)$ is also exact for the constant function and therefore $R_n[B] = R_n[B_1]$. We have $B_k \subset B_1$, $k \geqslant 2$. Let $f \in B_1$, then

$$g(x) = \frac{1}{n} \sum_{i=1}^{n} f\left(x + \frac{2\pi i}{n}\right) \in B_n$$

and therefore

$$\begin{aligned}
\mathscr{L}_n(g) &= \frac{2\pi}{n} \sum_{i=1}^{n} \sum_{j=1}^{m} A_j \frac{1}{n} \sum_{s=1}^{n} f\left(\frac{2\pi(i-1)}{n} + \alpha_j \frac{2\pi}{n} + \frac{2\pi s}{n}\right) \\
&= \frac{2\pi}{n^2} \sum_{i=1}^{n} \sum_{j=1}^{m} \sum_{s=1}^{n} A_j f\left(\frac{2\pi(i+s-1)}{n} + \alpha_j \frac{2\pi}{n}\right) \\
&= \frac{2\pi}{n} \sum_{i=1}^{n} \sum_{j=1}^{m} A_j f\left(\frac{2\pi(i-1)}{n} + \alpha_j \frac{2\pi}{n}\right) \\
&= \mathscr{L}_n(f)
\end{aligned}$$ (3.49)

since from the 2π-periodicity of f it follows that for every integer i we have

$$\sum_{s=1}^{n} f\left(\frac{2\pi(i+s-1)}{n} + \alpha_j \frac{2\pi}{n}\right) = \sum_{s=1}^{n} f\left(\frac{2\pi(s-1)}{n} + \alpha_j \frac{\pi}{n}\right)$$

Since

$$\int_0^{2\pi} f(x)\,dx = \int_0^{2\pi} g(x)\,dx = 0$$

it follows from (3.49) that $R_n(g) = R_n(f)$. Since $f \in B_1$ was arbitrary, we obtain

$$R_n[B_n] \geqslant R_n[B_1] = R_n[B]$$

Since $B_n \subset B_1$,

$$\begin{aligned}
R_n[B_n] &= R_n[B_1] \\
&= \sup\left\{ \left| \frac{2\pi}{n} \sum_{i=1}^{n} \sum_{j=1}^{m} A_j f\left(\frac{2\pi(i-1)}{n} + \alpha_j \frac{2\pi}{n}\right) \right| : f \in B_n \right\} \\
&= \sup\left\{ \left| \frac{2\pi}{n} \sum_{i=1}^{n} \sum_{j=1}^{m} A_j f\left(\alpha_j \frac{2\pi}{n}\right) \right| : f \in B_n \right\} \\
&\leqslant 2\pi\mu_n[B] \sum_{j=1}^{n} |A_j| \qquad\qquad \blacksquare
\end{aligned}$$

The following statement is well known (see Nikolskii [1] p. 218).

Lemma 3.3. *For every* $r = 2, 4, 6, \ldots$ *we have*

$$\mu_n[\tilde{W}^r L(1; 0, 2\pi)] = \frac{K_{r-1}}{4n^r}$$

where

$$K_r = \frac{4}{\pi} \sum_{v=0}^{\infty} (-1)^{v(r+1)} (2v+1)^{-r-1}$$

are the Favard constants.

From Lemmas 3.2 and 3.3 we derive the following theorem.

Theorem 3.10. *Let* f *be a* 2π-*periodic function with* $f^{(r-1)}$ *absolutely continuous for* $r = 2, 4, 6, \ldots$. *Then*

$$|R_n(f)| \leqslant 2\pi \sum_{j=1}^{m} \frac{K_{r-1}}{4n^r} |A_j| \|f^{(r)}\|_{L[0, 2\pi]}$$

Theorem 3.11. *Let* f *be a bounded* 2π-*periodic integrable function. Then for the error of the nth composite quadrature formula* (3.48), *for every* $r = 1, 2, \ldots$, *we have the estimate:*

$$|R_n(f)| \leqslant c(r) \left(1 + \sum_{j=1}^{m} |A_j| \right) \tau_r(f; n^{-1})_L$$

where $c(r)$ *are constants, dependent only on* r.

Proof. Since $\tau_k(f; \delta)_L \leqslant 2^{2k+3} \tau_{k-1}(f; \delta)_L$, it is clear that it is sufficient to prove Theorem 3.11 only for even r. We shall use Theorem 2.5. This theorem states that for every even r and $h = 1/n$ there exists a function f_r which we can assume to be 2π-periodic (since this is true for f, see the proof of Theorem 2.5 in the periodic case), with the following properties:

(i) $|f(x) - f_r(x)| \leqslant \omega_r \left(f, x; \frac{2}{n} \right)$

(ii) $\|f - f_r\|_{L[0, 2\pi]} \leqslant \omega_r(f; n^{-1})_{L[0, 2\pi]}$

(iii) $f_r \in \tilde{W}^r L, \|f_r^{(r)}\|_{L[0, 2\pi]} \leqslant c(r) n^r \omega_r(f; n^{-1})_L$,

where $c(r)$ is a constant, dependent only on r.

We have

$$|R_n(f)| \leqslant |R_n(f - f_r))| + |R_n(f_r)| \tag{3.50}$$

For $|R_n(f_r)|$, using Theorem 3.10 and property (iii), we obtain

$$|R_n(f_r)| \leqslant 2\pi \sum_{j=1}^{m} |A_j| \frac{K_{r-1}}{4n^r} c(r) n^r \omega_r(f; n^{-1})_L$$

$$\leqslant c(r) \frac{\pi}{2} K_{r-1} \sum_{j=1}^{m} |A_j| \tau_r(f; n^{-1})_{L[0, 2\pi]} \tag{3.51}$$

For $|R_n(f - f_r)|$ we obtain, using (i), (ii) and (3.48),

$$|R_n(f - f_r)| \leqslant \int_0^{2\pi} |f(x) - f_r(x)| dx$$

$$+ \sum_{i=1}^{n} \sum_{j=1}^{m} \frac{2\pi}{n} |A_j| f[x_{i-1} + \alpha_j(x_i - x_{i-1})]$$

$$- f_r[x_{i-1} + \alpha_j(x_i - x_{i-1})]$$

$$\leqslant \omega_r(f; n^{-1})_{L[0,2\pi]} + \sum_{i=1}^{n} \sum_{j=1}^{m} \frac{2\pi}{n} |A_j| \omega_r\left(f, x_{i-1} + \alpha_j \frac{2\pi}{n}; \frac{2}{n} \right)$$

$$\leqslant \tau_r(f; n^{-1})_{L[0,2\pi]} + \sum_{i=1}^{n} \frac{2\pi}{n} \omega_r\left(f, x_{i-1}; \frac{4\pi}{n} \right) \sum_{j=1}^{n} |A_j| \qquad (3.52)$$

From Lemma 2.5 it follows, that

$$\sum_{i=1}^{n} \frac{2\pi}{n} \omega_r\left(f, x_{i-1}; \frac{4\pi}{n} \right) \leqslant c'(r)\tau_r(f; n^{-1})_{L[0,2\pi]}$$

and therefore (3.52) gives us

$$|R_n(f - f_r)| \leqslant \left(1 + c'(r) \sum_{j=1}^{m} |A_j| \right) \tau_r(f; n^{-1})_{L[0,2\pi]} \qquad (3.53)$$

where $c'(r)$ is a constant, dependent only on r.

The Theorem now follows from inequalities (3.50), (3.51) and (3.52). ∎

Corollary 3.1. If f is a 2π-periodic absolutely continuous function, then

$$|R_n(f)| = O(n^{-1}\omega_r(f'; n^{-1})_L)$$

where the constant in O depends only on r and $\sum_{i=1}^{m} |A_i|$.

Notice that, using the interpolation method, we again obtain large constants. It is possible to obtain better constants if we work directly with concrete quadratures for 2π-periodic functions. Since our aim is only to obtain quantitative effects connected with the estimates via the averaged moduli of smoothness, we shall not consider here the problem of deriving exact constants.

Remark. For 2π-periodic functions, the local modulus of kth order is given by

$$\omega_k(f, x; \delta) = \sup\left\{ |\Delta_h^k f(t)| : t, t + kh \in \left[x - \frac{k\delta}{2}, x + \frac{k\delta}{2} \right] \right\}$$

and the averaged modulus of smoothness of kth order is

$$\tau_k(f; \delta)_p = \tau_k(f; \delta)_{L_p[0,2\pi]}$$
$$= \| \omega_k(f, \cdot; \delta) \|_{L_p[0,2\pi]}$$

3.5. Notes

The first application of the averaged moduli of smoothness to the Newton–Cotes composite quadrature formulae (for example Simpson's rule and trapezoidal rule) was due to Popov [4], [5]. All results in Sections 3.2, 3.3 and 3.4 were derived by Ivanov [1].

In his doctoral thesis Proinov applies the averaged modulus of smoothness to general quadrature processes. Let

$$X = (x_i^{(n)}), 0 \leqslant x_1^{(n)} \leqslant \cdots \leqslant x_n^{(n)} \leqslant 1 \text{ and } P = (p_i^{(n)}), p_i^{(n)} \geqslant 0, i = 1, \ldots, n, \sum_{i=1}^{n} p_i^{(n)} = 1,$$

$n = 1, 2, \ldots$, be two triangular matrices. The following quadrature process is considered:

$$\int_0^1 f(x)\,dx = \sum_{i=1}^{n} p_i^{(n)} f(x_i^{(n)}) + R_n(f) \quad n = 1, 2, \ldots$$

For such a quadrature process the following characteristic (called discrepance) is useful:

$$D_n = D_n(X, P) = \sup \left\{ \left| \sum_{i=1}^{n} p_i^{(n)} \lambda_\alpha(x_i^{(n)}) - \alpha \right| ; 0 \leqslant \alpha \leqslant 1 \right\}$$

where $\lambda_\alpha(x)$ is the characteristic function of the interval $[0, \alpha)$. First Niederreiter [1] obtained an estimate for the error $|R_n(f)|$ using $\omega(f; D_n)$, where ω is the uniform modulus of continuity of the function f. The following result is that of Proinov [1]:

$$|R_n(f)| \leqslant \tau(f; D_n)_1$$

There are some generalizations of this result (see, for example, Totkov and Baselkov [1], Hristov [1] has obtained estimations for the quadrature process, which include derivatives of the function f, using $\tau(f^{(r)}; D_n)$ where the process uses up to the rth derivative of f.

(See also Totkov [1].)

4 APPROXIMATION OF FUNCTIONS BY MEANS OF LINEAR SUMMATION OPERATORS IN L_p

In this chapter we shall consider approximation of bounded measurable functions f in the interval $[a, b]$ by means of linear summation operators of the type

$$L_n(f; x) = \sum_{i=0}^{n} f(x_i)\varphi_i(x) \tag{4.1}$$

where $a \leqslant x_0 < x_1 < \cdots < x_n \leqslant b$ are $n + 1$ fixed points in the interval $[a, b]$, and the functions φ_i, $i = 0, \ldots, n$, are in $L_p[a, b]$.

To the class of operators of type (4.1) belong all interpolation operators (independent of whether φ_i are algebraic polynomials, trigonometric polynomials, splines or other basic functions) and many of the often used linear positive operators such as the classical Bernstein operators, Szasz–Mirakian operators, Baskakov's operators, etc. Usually for operators of type (4.1) uniform estimates are known, i.e. estimates for the uniform distance $\| L(f; \cdot) - f \|_{C[a,b]}$, and the estimates use $\omega_1(f; \delta)$ or $\omega_2(f; \delta)$. Our aim is to obtain estimates for $\| L_n(f; \cdot) - f \|_{L_p[a,b]}$ using $\tau_k(f; \delta)_p$. From these, when $p = \infty$ the known estimates will follow. Let us note that for operators of type (4.1) estimates using $\omega_k(f; \delta)_p$ are not valid, since $L_n(f; x)$ depend only on the values of the function f at the points x_i, $i = 0, 1, \ldots, n$, but the integral moduli $\omega_k(f; \delta)_p$ do not depend on the values of the function at a finite number of points.

4.1. Estimates for Bernstein operators and Szasz–Mirakian operators

The Bernstein operators and Szasz–Mirakian operators are classical examples of linear positive operators. For every function f, bounded in the interval $[0, 1]$, the Bernstein polynomial of nth degree (Bernstein operator of order n) is given by

$$B_n(f;x) = \sum_{k=0}^{n} f\left(\frac{k}{n}\right)\binom{n}{k}x^k(1-x)^{n-k} \tag{4.2}$$

For every function f bounded on $[0, \infty)$ the Szasz–Mirakian operator of order n is given by

$$M_n(f;x) = e^{-nx} \sum_{k=0}^{\infty} f\left(\frac{k}{n}\right)\frac{(nx)^k}{k!} \tag{4.3}$$

In this chapter we shall estimate the L_p-distance for $1 \leqslant p \leqslant \infty$ between the function f and $B_n(f;\cdot)$, $M_n(f;\cdot)$ using the moduli $\tau_2(f;\delta)_p$ (see Andreev and Popov [1]).

The following classical results are well-known.

Let f be a continuous function in $[0, 1]$. Then

$$\|B_n(f;\cdot) - f\|_{C[0,1]} = O[\omega(f;n^{-1/2})] \tag{4.4}$$

(Popoviciu's theorem).

Let f be a continuous function in the interval $[0, \infty]$, $f(x) = 0$ for $x \geqslant 0$, then

$$\|M_n(f;\cdot) - f\|_{C[0, a]} = O[\omega(f;n^{-1/2})] \tag{4.5}$$

To obtain our estimates we shall need the following lemmas.

Lemma 4.1. *Let f be a bounded function in the interval $[0, 1]$. Then*

$$\|B_n(f;\cdot)\|_{L_p[0,1]} \leqslant \|f\|_{l_{\Sigma_n}^p}$$

where

$$\Sigma_n = \left\{\frac{i}{n}; i = 0, 1, \ldots, n\right\}$$

$$\|f\|_{l_{\Sigma_n}^p} = \left\{\sum_{i=0}^{n}\left|f\left(\frac{i}{n}\right)\right|^p n^{-1}\right\}^{1/p}$$

Proof. We shall use Jensen's inequality:

$$\left|\sum_{i=0}^{n}\alpha_i\beta_i\right|^p \leqslant \sum_{i=0}^{n}\alpha_i|\beta_i|^p \quad \alpha_i \geqslant 0, \sum_{i=1}^{n}\alpha_i = 1, p \geqslant 1 \tag{4.6}$$

We have

$$\|B_n(f;\cdot)\|_p = \left[\int_0^1\left|\sum_{k=0}^{n} f\left(\frac{k}{n}\right)\binom{n}{k}x^k(1-x)^{n-k}\right|^p dx\right]^{1/p}$$

$$\leqslant \left[\int_0^1\sum_{k=0}^{n}\binom{n}{k}x^k(1-x)^{n-k}\left|f\left(\frac{k}{n}\right)\right|^p dx\right]^{1/p}$$

$$\leqslant \left[\sum_{k=0}^{n}\left|f\left(\frac{k}{n}\right)\right|^p\int_0^1\binom{n}{k}x^k(1-x)^{n-k} dx\right]^{1/p}$$

$$= \left[\sum_{k=0}^{n} \left| f\left(\frac{k}{n}\right) \right|^p (n+1)^{-1} \right]^{1/p}$$

$$\leq \left[\sum_{k=0}^{n} \left| f\left(\frac{k}{n}\right) \right|^p n^{-1} \right]^{1/p} = \| f \|_{\ell_n^p} \qquad \blacksquare$$

Lemma 4.2. *Let f be a bounded function in the interval $[0, a], f(x) = 0, x \geq a > 0$. Then*

$$\| M_n(f; \cdot) \|_{L_p[0,a]} \leq \| f \|_{\ell_n^p}$$

where

$$\Sigma_n = \left\{ \frac{i}{n}; i = 0, 1, \ldots, k; \frac{k}{n} \leq a < \frac{k+1}{n} \right\}$$

$$\| f \|_{\ell_n^p} = \left[\sum_{i=0}^{k} \left| f\left(\frac{i}{n}\right) \right|^p n^{-1} \right]^{1/p}$$

Proof. Again, using Jensen's inequality (4.6), we obtain

$$\| M_n(f; \cdot) \|_{L_p[0,a]} = \left[\int_0^a \left| \sum_{i=0}^{k} f\left(\frac{i}{n}\right) e^{-nx} \frac{(nx)^i}{i!} \right|^p dx \right]^{1/p}$$

$$\leq \left[\sum_{i=0}^{k} \left| f\left(\frac{i}{n}\right) \right|^p \int_0^a e^{-nx} \frac{(nx)^i}{i!} dx \right]^{1/p}$$

$$\leq \left\{ \sum_{i=0}^{k} \left| f\left(\frac{i}{n}\right) \right|^p \int_0^\infty e^{-nx} \frac{(nx)^i}{i!} dx \right\}^{1/p} = \| f \|_{\ell_n^p}$$

since

$$\int_0^\infty e^{-nx} \frac{(nx)^i}{i!} dx = \frac{1}{n} \qquad \blacksquare$$

Lemma 4.3. *If $f \in W_p^2[0, 1]$, then*

$$\| B_n(f; \cdot) - f \|_{L_p[0,1]} \leq c \| f'' \|_{L_p[0,1]} n^{-1}$$

where c is an absolute constant.

Proof. From Taylor's formula for $x, t \in [0, 1]$, we have

$$f(t) = f(x) + (t - x) f'(x) + \int_x^1 (t - \theta)_+ f''(\theta) d\theta + \int_0^x (\theta - t)_+ f''(\theta) d\theta \qquad (4.7)$$

where

$$t_+ = \begin{cases} 0 & t \leq 0 \\ t & t \geq 0 \end{cases}$$

In fact, if $t > x$ in (4.7), we have only the first integral, and if $t < x$ we have only the second integral. we now integrate by parts.

Using the fact that B_n is a linear summation operator and that $B_n(t; x) = x$ (see Natarson [1]), we obtain from (4.7)

$$B_n(f; x) - f(x) = \int_x^1 B_n((t - \theta)_+; x) f''(\theta) \, d\theta + \int_0^x B_n((\theta - t)_+; x) f''(\theta) \, d\theta$$

$$= \int_0^1 K(x; \theta) f''(\theta) \, d\theta, \qquad (4.8)$$

where

$$K(x; \theta) = \begin{cases} B_n((t - \theta)_+; x) & \theta \geqslant x \\ B_n((\theta - t)_+; x) & \theta < x \end{cases}$$

Therefore

$$\| B_n(f; \cdot) - f \|_{L[0,1]} \leqslant \int_0^1 \int_0^1 |K(x; \theta)| |f''(\theta)| \, d\theta \, dx$$

$$\leqslant \| f'' \|_{L[0,1]} \max_{0 \leqslant \theta \leqslant 1} \int_0^1 |K(x, \theta)| \, dx \qquad (4.9)$$

Let us now estimate $\int_0^1 K(x; \theta) \, dx$ using the notation

$$p_{n,k}(x) = \binom{n}{k} x^k (1 - x)^{n-k}$$

Integrating by parts it is easy to prove the following:

$$\int_0^\theta p_{n,k}(x) \, dx = \frac{1}{n+1} \sum_{i=k+1}^{n+1} p_{n+1,i}(\theta)$$

$$\int_0^1 p_{n,k}(x) \, dx = \frac{1}{n+1} \sum_{i=0}^{k} p_{n+1,i}(\theta)$$

Using these equations we obtain

$$\int_0^1 |K(x; \theta)| \, dx = \int_0^\theta B_n((t - \theta)_+; x) \, dx + \int_\theta^1 B_n((\theta - t)_+; x) \, dx$$

$$= \sum_{k=0}^{n} \left[\left(\frac{k}{n} - \theta \right)_+ \int_0^\theta p_{n,k}(x) \, dx + \left(\theta - \frac{k}{n} \right)_+ \int_\theta^1 p_{n,k}(x) \, dx \right]$$

$$= \frac{1}{n+1} \sum_{k=0}^{n} \left[\left(\frac{k}{n} - \theta \right)_+ \sum_{i=k+1}^{n+1} p_{n+1,i}(\theta) + \left(\theta - \frac{k}{n} \right)_+ \sum_{i=0}^{k} p_{n+1,i}(\theta) \right]$$

$$(4.10)$$

Let us set

$$s_k(\theta) = \frac{1}{n+1} \sum_{i=0}^{k} \left(\frac{i}{n} - \theta\right)_+, \quad s_{-1}(\theta) = 0$$

$$d_k(\theta) = \frac{1}{n+1} \sum_{i=k}^{n} \left(\theta - \frac{i}{n}\right)_+, \quad d_{n+1}(\theta) = 0 \tag{4.11}$$

From (4.11), on changing the order of summation we obtain

$$\frac{1}{n+1} \sum_{k=0}^{n} \left(\frac{k}{n} - \theta\right)_+ \sum_{i=k+1}^{n+1} P_{n+1,i}(\theta) = \sum_{k=0}^{n} s_k(\theta) P_{n+1,k+1}(\theta)$$

$$\frac{1}{n+1} \sum_{k=0}^{n} \left(\theta - \frac{k}{n}\right)_+ \sum_{i=0}^{k} P_{n+1,i}(\theta) = \sum_{k=0}^{n} d_k(\theta) P_{n+1,k}(\theta) \tag{4.12}$$

We shall now estimate s_k and d_k from above.
For $k \leqslant n\theta$ we have $s_k(\theta) = 0$. If $\theta < k/n$, then

$$\frac{1}{n+1} \sum_{i=0}^{k} \left(\frac{i}{n} - \theta\right)_+ = \frac{1}{n+1} \sum_{i=[\theta n]+1}^{k} \left(\frac{i}{n} - \theta\right)$$

$$= \frac{1}{n(n+1)} \left[\frac{k(k+1)}{2} - \frac{[n\theta]([n\theta]+1)}{2}\right] - \frac{\theta}{n+1}(k - [n\theta])$$

$$\leqslant \frac{1}{2} \left(\frac{k+1}{n+1} - \theta\right)^2 + \frac{3}{2n} \tag{4.13}$$

In an analogous way we find that

$$d_k(\theta) = 0 \quad k \geqslant n\theta$$

$$d_k(\theta) \leqslant \frac{1}{2} \left(\frac{k}{n+1} - \theta\right)^2 + \frac{3}{2n} \quad k < n\theta \tag{4.14}$$

From (4.10)–(4.14) we obtain

$$\int_0^1 |K(x;\theta)| dx \leqslant \sum_{k=0}^{n} s_k(\theta) P_{n+1,k+1}(\theta) + \sum_{k=0}^{n} d_k(\theta) P_{n+1,k}(\theta)$$

$$= \sum_{k=0}^{n} s_{k-1}(\theta) P_{n+1,k}(\theta) + \sum_{k=1}^{n} d_k(\theta) P_{n+1,k}(\theta)$$

$$+ s_n(\theta) P_{n+1,n+1}(\theta) + d_0(\theta) P_{n+1,0}(\theta)$$

$$= \sum_{k=0}^{n+1} [s_{k-1}(\theta) + d_k(\theta)] P_{n+1,k}(\theta)$$

$$\leqslant \sum_{k=0}^{n+1} \left[\frac{1}{2} \left(\frac{k}{n+1} - \theta\right)^2 + \frac{3}{2n}\right] P_{n+1,k}(\theta) \tag{4.15}$$

Using the well-known results (see, for example, Natarson [1])

$$\sum_{k=0}^{n+1} p_{n+1,k}(x) = 1$$

$$\sum_{k=0}^{n+1} \frac{k}{n+1} p_{n+1,k}(x) = x$$

$$\sum_{k=0}^{n+1} \frac{k^2}{(n+1)^2} p_{n+1,k}(x) = x^2 + \frac{x(1-x)}{n+1}$$

we obtain from (4.15)

$$\int_0^1 |K(x;\theta)|\,dx \leqslant \frac{\theta(1-\theta)}{2(n+1)} + \frac{3}{2n} \leqslant \frac{13}{8n} \tag{4.16}$$

From (4.9) and (4.16) it follows that

$$\| B_n(f;\cdot) - f \|_{L[0,1]} \leqslant \frac{13}{8} \frac{\| f'' \|_{L[0,1]}}{n} \tag{4.17}$$

On the other hand, from (4.7) we obtain

$$|B_n(f;x) - f(x)| \leqslant B_n\left(\left[\int_x^1 (t-\theta)_+ |f''(\theta)|\,d\theta + \int_0^x (\theta-t)_+ |f''(\theta)|\,d\theta\right];x\right)$$

$$\leqslant \| f'' \|_{C[0,1]} B_n\left(\int_x^t (t-\theta)_+ \,d\theta + \int_t^x (\theta-t)_+ \,d\theta; x\right)$$

$$\leqslant \| f'' \|_{C[0,1]} B_n(\tfrac{1}{2}(t-x)^2; x)$$

$$= \frac{x(1-x)}{2n} \| f'' \|_c \leqslant \frac{\| f'' \|_c}{8n} \tag{4.18}$$

The inequalities (4.17) and (4.18) together with the interpolation Theorem 2.8 give us, for every p, $1 \leqslant p \leqslant \infty$, the inequality

$$\| B_n(f;\cdot) - f \|_{L_p[0,1]} \leqslant c \frac{\| f'' \|_{L_p[0,1]}}{n} \tag{4.19}$$

∎

Lemma 4.4. *Let f be bounded in $[0,\infty)$, $f \in W_p^2[0,\infty)$, $1 \leqslant p \leqslant a$, $f(x) = 0$ for $x \geqslant a > 0$, then the following estimate holds:*

$$\| M_n(f;\cdot) - f \|_{L_p[0,a]} \leqslant c(a)n^{-1} \| f'' \|_{L_p[0,a]}$$

where $c(a)$ is a constant, dependent only on a.

Proof. In an analogous way to (4.9), using Taylor's formula on $[0,\infty)$ and the conditions of the lemma, we obtain

$$\| M_n(f;\cdot) - f \|_{L_p[0,a]} \leqslant a^{1/p} \| f'' \|_{L_p[0,a]} \max_{0 \leqslant \theta \leqslant a} \int_0^a N(x;\theta)\,dx \tag{4.20}$$

where

$$N(x; \theta) = \begin{cases} M_n((t - \theta)_+; x) & \text{for } \theta \geqslant x \\ M_n((\theta - t)_+; x) & \text{for } \theta < x \end{cases}$$

we use the notation

$$q_{n,k}(x) = e^{-nx} \frac{(nx)^k}{k!}$$

$$r_k(\theta) = \frac{1}{n} \sum_{i=0}^{k} \left(\frac{i}{n} - \theta \right)_+, \quad r_{-1}(\theta) = 0 \tag{4.21}$$

$$t_k(\theta) = \frac{1}{n} \sum_{i=k}^{\infty} \left(\theta - \frac{i}{n} \right)_+$$

As before we have:

$$\int_0^a N(x; \theta) \, dx = \int_0^\theta M_n((t - \theta)_+; x) \, dx + \int_\theta^a M_n((\theta - t)_+; x) \, dx$$

$$\leqslant \sum_{k=0}^{\infty} \left[\left(\frac{k}{n} - \theta \right)_+ \int_0^\theta q_{n,k}(x) \, dx + \left(\theta - \frac{k}{n} \right)_+ \int_\theta^\infty q_{n,k}(x) \, dx \right] \tag{4.22}$$

Integrating by parts we have the following equations

$$\int_0^\theta q_{n,k}(x) \, dx = \frac{1}{n} \sum_{i=k+1}^{\infty} q_{n,i}(\theta)$$

$$\int_\theta^\infty q_{n,k}(x) \, dx = \frac{1}{n} \sum_{i=0}^{k} q_{n,i}(\theta)$$

From these equations and (4.22), using the Abel transformation and the notation of (4.21), we obtain

$$\int_0^a N(x; \theta) \, dx \leqslant \frac{1}{n} \sum_{k=0}^{\infty} \left[\left(\frac{k}{n} - \theta \right)_+ \sum_{i=k+1}^{\infty} q_{n,i}(\theta) + \left(\theta - \frac{k}{n} \right)_+ \sum_{i=0}^{k} q_{n,i}(\theta) \right]$$

$$= \sum_{k=0}^{\infty} \left\{ [r_k(\theta) - r_{k-1}(\theta)] \sum_{i=k+1}^{\infty} q_{n,i}(\theta) + [t_k(\theta) - t_{k-1}(\theta)] \sum_{i=0}^{k} q_{n,i}(\theta) \right\}$$

$$= \sum_{k=0}^{\infty} r_k(\theta) q_{n,k+1}(\theta) + \sum_{k=0}^{\infty} t_k(\theta) q_{n,k}(\theta) \tag{4.23}$$

As in the case of the Bernstein polynomials we see that

$$r_k(\theta) = 0 \quad \text{for } k \leqslant n\theta, \quad r_k(\theta) \leqslant \left(\frac{k+1}{n} - \theta \right)^2 \quad \text{for } k > n\theta$$

$$t_k(\theta) = 0 \quad \text{for } k \geqslant n\theta, \quad t_k(\theta) \leqslant \left(\frac{k}{n} - \theta \right)^2 + \frac{1}{n^2} \quad \text{for } k < n\theta \tag{4.24}$$

From (4.23) and (4.24) we obtain

$$\int_0^a N(x,\theta)\,dx \leqslant \sum_{k=0}^{\infty}(r_{k-1}(\theta)+t_k(\theta))q_{n,k}(\theta)$$

$$\leqslant \sum_{k=0}^{\infty}\left[\left(\frac{k}{n}-\theta\right)^2+\frac{1}{n^2}\right]q_{n,k}(\theta) \leqslant \frac{a+1}{n} \tag{4.25}$$

since

$$\sum_{k=1}^{\infty}q_{n,k}(x)=1, \quad \sum_{k=0}^{\infty}\frac{k}{n}q_{n,k}(x)=x$$

$$\sum_{k=0}^{\infty}\frac{k^2}{n^2}q_{n,k}(x)=x^2+\frac{x}{n} \quad 0\leqslant\theta\leqslant a \tag{4.26}$$

From (4.20) and (4.25) we obtain

$$\|M_n(f;\cdot)-f\|_{L[0,a]} \leqslant \frac{a(a+1)}{n}\|f''\|_{L[0,a]} \tag{4.27}$$

Using (4.26) in the same way as (4.18), we obtain

$$\|M_n(f;\cdot)-f\|_{C[0,a]} \leqslant \frac{a}{2n}\|f''\|_{C[0,a]} \tag{4.28}$$

Using the interpolation Theorem 2.8, from (4.27) and (4.28) we find that

$$\|M_n(f;\cdot)-f\|_{L_p[0,a]} \leqslant c\,\frac{a(a+1)^{1/p}}{n}\|f''\|_{L_p[0,a]} \tag{4.29}$$

∎

Since we have Lemmas 4.1 and 4.3 and

$$\|f\|_{\ell_n^p} \leqslant \left[\sum_{i=0}^{n}\left|f\left(\frac{i}{n}\right)\right|^p\frac{2}{n}\right]^{1/2}$$

we can apply Theorem 2.9 to Bernstein operators.

Theorem 4.1. *Let f be a bounded measurable function in* $[0,1]$, *then*

$$\|B_n(f;\cdot)-f\|_{L_p[0,1]} \leqslant c\tau_2(f;n^{-1/2})_{L_p}$$

where c is an absolute constant.

Since in Lemmas 4.2 and 4.4 there is the additional condition $f(x)=0$ for $x\geqslant a$, we cannot directly apply the interpolation Theorem 2.8 to the Szasz–Mirakian operators. We see immediately, however, that if we use the intermediate function

$$f_h(x)=-\frac{1}{h^2}\int_0^h\int_0^h\left\{f(x+t_1+t_2)-2f[x+\tfrac{1}{2}(t_1+t_2)]\right\}dt_1\,dt_2$$

for which it is obvious that $f_h(x)=0$, $f_h''(x)=0$ for $x\geqslant a$ if $f(x)=0$ for $x\geqslant a$, then the proof of the theorem remains valid and therefore Lemmas 4.2 and 4.4 give us the following.

Theorem 4.2. *Let f be a bounded measurable function on $[0, \infty)$ and $f(x) = 0$ for $x \geq a$. Then*

$$\| M_n(f; \cdot) - f \|_{L_p[0,a]} \leq c(a)\tau_2(f; n^{-1/2})_p$$

where the constant $c(a)$ depends only on a (more precisely $c(a) = c'a(a + 1)^{1/p}$, where c' is an absolute constant).

From Theorems 4.1 and 4.2 we obtain, using the properties of the modulus $\tau_2(f; \delta)_p$:

Corollory 4.1. *Let f be a function of bounded variation in the interval $[0, 1]$ (respectively $[0, \infty)$) with $f(x) = 0$ for $x \geq a$. Then*

$$\| B_n(f; \cdot) - f \|_{L[0,1]} \leq c_1 V_0^1 f n^{-1/2}$$
$$\| M_n(f; \cdot) - f \|_{L[0,a]} \leq c_2(a) V_0^a f n^{-1/2}$$

Corollary 4.2. *Let f be an absolutely continuous function in the interval $[0, 1]$ (respectively $[0, \infty)$), with $f' \in L_p[0, 1]$ (respectively $f' \in L_p[0, \infty)$) and $f(x) = 0$ for $x \geq a$. Then*

$$\| B_n(f; \cdot) - f \|_{L_p[0,1]} \leq c_3 n^{-1/2} \omega(f'; n^{-1/2})_{L_p[0,1]}$$
$$\| M_n(f; \cdot) - f \|_{L_p[0,a]} \leq c_4(a) n^{-1/2} \omega(f'; n^{-1/2})_{L_p[0,a]}$$

Therefore, if f is an absolutely continuous function in $[0, 1]$ then

$$\| B_n(f; \cdot) - f \|_{L[0,1]} = o(n^{-1/2})$$

Corollary 4.3. *Let f' be a function of bounded variation in the interval $[0, 1]$ (respectively $[0, \infty)$) with $f(x) = 0$ for $x \geq a$. Then*

$$\| B_n(f; \cdot) - f \|_{L[0,1]} = c_5 n^{-1} V_0^1 f'$$
$$\| M_n(f; \cdot) - f \|_{L[0,a]} \leq c_6 n^{-1} V_0^a f'$$

Hoeffding [1] has the following result for the Bernstein polynomials: If $V_0^1 f < \infty$, then

$$\int_0^1 |B_n(f; x) - f(x)| dx \leq c_n \mathcal{T}(f) + V_0^1 f(n + 1)^{-1}$$

where

$$c_n = \sqrt{\frac{2}{en}}, \quad \mathcal{T}(f) = \int_0^1 \sqrt{x(1 - x)} |df|$$

Obviously Corollary 4.1 for the Bernstein polynomials also follows from this result.

The technique developed here can be applied also to other linear summation operators, for example to the well-known Baskakov operators, which are a generalization of the Szasz–Mirakian operators (see Djukanova [1]). We shall not present these estimates here.

4.2. Korovkin's theorem in L_p

The estimates in the previous section were obtained using the concrete form of the Bernstein and Szasz–Mirakian operators. In the general case of approximation of functions by means of linear positive operators in L_p, it is also possible to obtain a generalization of the well-known Korovkin theorem, more exactly a generalization of its quantitative version, due to Mamedov [1] (see Korovkin [1]). Let us first recall Korovkin's theorem.

Korovkin's Theorem. *Let L be a linear positive operator in the space $C[a, b]$, i.e. $L:C[a, b] \to C[a, b]$, and*

$$L(1; x) = 1, \quad L(t; x) = x + \alpha(x), \quad L(t^2; x) = x^2 + \beta(x) \qquad (4.30)$$

Then for every function $f \in C[a, b]$ we have

$$\| L(f; \cdot) - f \|_{C[a,b]} \leqslant 3\omega(f; \sqrt{d})$$

where

$$d = \| \beta(x) - 2x\alpha(x) \|_{C[a,b]} \qquad (4.31)$$

Our aim will be to obtain an estimate for $\| Lf - f \|_{L_p[a,b]}$ by $\tau(f; \delta)_p$, $1 \leqslant p < \infty$.

Lemma 4.5. *Let*

$$\sigma(c; x) = \begin{cases} 0 & x < c \\ 1 & x \geqslant c \end{cases}$$

and let L be a linear positive operator defined in the space $M[a, b]$ of all bounded measurable functions in $[a, b]$, i.e. $L: M[a, b] \to M[a, b]$. Let L satisfy the condition (4.30). Then for every $c \in (a, b)$ the following estimates hold:

(a) $\| L(\sigma(c; t); \cdot) - \sigma(c; \cdot) \|_{L[a,b]} \leqslant 4\sqrt{d}$

(b) $|L(\sigma(c; t); x) - \sigma(c; x)| \leqslant \dfrac{d}{(x - c)^2}$ *for $x \in [a, b]$, $x \neq c$*

where d is given by (4.31).

Proof. Let $x_0 \in [a, c)$. Then the quadratic $\varphi_{x_0}(x) = \theta(x - x_0)^2$, for $\theta = 1/(c - x_0)^2$, satisfies

$$0 \leqslant \sigma(c; x) \leqslant \varphi_{x_0}(x) \quad x \in [a, b]$$

and therefore, since L is a linear positive operator satisfying the conditions

(4.30), we have

$$0 \leqslant L(\sigma(c;t);x_0)$$

$$\leqslant L(\varphi_{x_0}(t);x_0)$$

$$= \frac{1}{(c-x_0)^2} L((t-x_0)^2;x_0)$$

$$= \frac{\beta(x_0) - 2x_0\alpha(x_0)}{(c-x_0)^2}$$

i.e.

$$0 \leqslant L(\sigma(c;t);x_0) \leqslant \frac{d}{(c-x_0)^2} \quad x_0 \in [a,c] \qquad (4.32)$$

We now integrate this inequality from a to $c-h$, $h>0$, giving

$$\int_a^{c-h} L(\sigma(c;t);x)\,dx \leqslant d\left(\frac{1}{h} - \frac{1}{c-a}\right) \leqslant \frac{d}{h} \qquad (4.33)$$

In the same way, for $x_0 \in (c,b]$, the quadratic

$$\varphi_{x_0}(x) = 1 - \frac{(x-x_0)^2}{(x_0-c)^2}$$

for every $x \in [a,b]$, satisfies the inequalities

$$\varphi_{x_0}(x) \leqslant \sigma(c;x) \leqslant 1$$

Again, since L is a linear positive operator with the properties (4.30), we have

$$1 \geqslant L(\sigma(c;t);x_0)$$

$$\geqslant L(\varphi_{x_0}(t);x_0)$$

$$= 1 - \frac{1}{(x_0-c)^2} L((t-x_0)^2;x_0)$$

$$= 1 - \frac{\beta(x_0) - 2x_0\alpha(x_0)}{(x_0-c)^2}$$

i.e.

$$\frac{d}{(x-c_0)^2} \geqslant 1 - L(\sigma(c;t);x_0) \geqslant 0 \qquad (4.34)$$

Integrating this inequality from $c+h$ to b, we obtain

$$0 \leqslant \int_{c+h}^b [1 - L(\sigma(c;t);x)]\,dx \leqslant d\left(\frac{1}{h} - \frac{1}{b-c}\right) \leqslant \frac{d}{h} \qquad (4.35)$$

Again since L is a linear positive operator and $L1 = 1$ we have, for every $x \in [a,b]$,

$$0 \leqslant L(\sigma(c;t);x) \leqslant 1 \qquad (4.36)$$

Using (4.33), (4.35) and (4.36) we obtain

$$\| L(\sigma(c;t);\cdot) - \sigma(c;\cdot)\|_{L[a,b]}$$

$$= \left\{ \int_a^{c-h} L(\sigma(c;t);x)\,dx \right.$$

$$+ \int_{c-h}^c L(\sigma(c;t);x)\,dx + \int_c^{c+h} [1 - L(\sigma(c;t);x)]\,dx$$

$$\left. + \int_{c+h}^b [1 - L(\sigma(c;t);x)]\,dx \right\}$$

$$\leqslant \frac{d}{h} + 2h + \frac{d}{h}$$

$$= 2\left(\frac{d}{h} + h\right)$$

In this inequality, if $c - h < a$, the first integral is zero and the second integral is from a to c. If $c + h > b$ the fourth integral is zero and the third integral is from c to b. Therefore the above inequality holds for every h. Letting $h = \sqrt{d}$, we obtain statement (a) of the lemma. Statement (b) follows immediately from (4.32) and (4.34). ∎

Theorem 4.3. *Let L be a linear positive operator, $L:M[a,b] \to M[a,b]$, with the properties (4.30). Then for every $f \in M[a,b]$ and every p, $1 \leqslant p \leqslant \infty$, we have*

$$\| f - Lf \|_{L_p[a,b]} \leqslant 748\tau(f;\sqrt{d})_p$$

where d is given by (4.31).

Proof. We can assume that $\sqrt{d} < b - a$. Let the integer m be such that $(b-a)/m \leqslant \sqrt{d} < (b-a)/(m-1)$. We then divide the interval $[a,b]$ into m equal intervals by the points $x_i = a + ih$, $i = 0,\ldots,m$, $h = (b-a)/m$. Let

$$F_i = \sup\{f(t):t \in [x_{i-1},x_i]\} \quad i = 1,2,\ldots,m$$
$$G_i = \inf\{f(t):t \in [x_{i-1},x_i]\} \quad i = 1,2,\ldots,m$$

Then the functions

$$F(x) = F_i \quad x \in [x_{i-1},x_i), i = 1,2,\ldots,m, F(b) = F(b-h/2)$$
$$G(x) = G_i \quad x \in [x_{i-1},x_i), i = 1,2,\ldots,m, G(b) = G(b-h/2)$$

satisfy

$$G(x) \leqslant f(x) \leqslant F(x) \quad x \in [a,b] \tag{4.37}$$

$$|F(x) - G(x)| \leqslant \omega\left(f,x;\frac{b-a}{m}\right) \tag{4.38}$$

From the last inequality we obtain

$$\| F - G \|_{L_p[a,b]} = \left[\int_a^b |F(x) - G(x)|^p \, dx \right]^{1/p}$$

$$\leqslant \left\{ \int_a^b \left[\omega\left(f, x; \frac{b-a}{m} \right) \right]^p dx \right\}^{1/p}$$

$$= \tau \left(f; \frac{b-a}{m} \right)_p \leqslant \tau(f; \sqrt{d})_p$$

i.e.

$$\| F - G \|_{L_p[a,b]} \leqslant \tau(f; \sqrt{d})_p \tag{4.39}$$

Since L is a linear positive operator, for (4.37) and (4.39) we obtain

$$\| Lf - f \|_p \leqslant \| Lf - LF \|_p + \| LF - F \|_p + \| F - f \|_p$$
$$\leqslant \| LG - LF \|_p + \| LF - F \|_p + \| F - G \|_p$$
$$\leqslant \| LG - G \|_p + \| G - F \|_p + \| F - LF \|_p + \| LF - F \|_p + \| F - G \|_p$$
$$\leqslant 2 \| LF - F \|_p + \| LG - G \|_p + 2\tau(f; \sqrt{d})_p \tag{4.40}$$

Let us now estimate $\| LF - F \|_p$. Using the notation of Lemma 4.5 and letting $F_0 = 0$, we have

$$F(x) = \sum_{i=1}^m (F_i - F_{i-1})\sigma(x_{i-1}; x) \tag{4.41}$$

from which we obtain

$$|L(F(t); x) - F(x)| \leqslant \sum_{i=1}^m |F_i - F_{i-1}| |L(\sigma(x_{i-1}; t); x) - \sigma(x_{i-1}; x)| \tag{4.42}$$

Let $x \in \Delta_{i_0} = [x_{i_0-1}, x_{i_0}]$. Using estimate (b) of Lemma 4.5 we find that

$$|\sigma(x_{i-1}; x) - L(\sigma(x_{i-1}; t); x)| \leqslant \frac{d}{(x - x_{i-1})^2} \quad i < i_0, i > i_0 + 1 \tag{4.43}$$

From (4.36) it follows that for $i = i_0, i_0 + 1$ we have

$$|\sigma(x_{i-1}; x) - L(\sigma(x_{i-1}; t); x)| \leqslant 1 \tag{4.44}$$

Let us now estimate

$$\psi(x) = \sum_{i=1}^{i_0-1} \frac{d}{(x - x_{i-1})^2} + \sum_{i=i_0+2}^m \frac{d}{(x - x_{i-1})^2}$$

We have

$$\psi(x) \leqslant d \left[\sum_{i=0}^{i_0-1} \frac{m^2}{(i_0 - 1 - i + 1)^2 (b - a)^2} + \sum_{i=i_0+2}^m \frac{m^2}{(i_0 - i + 1)^2 (b - a)^2} \right]$$

$$\leqslant 2 \frac{dm^2}{(b-a)^2} \sum_{i=1}^\infty \frac{1}{i^2} \leqslant 8 \frac{d(m-1)^2}{(b-a)^2} \frac{\pi^2}{6} \leqslant \frac{4}{3}\pi^2 \tag{4.45}$$

since $\sqrt{d} \leqslant (b - a)/(m - 1)$.

In (4.42) the summation is really from $i = 2$ to m since

$$L(\sigma(x_0; t); x) = L(1; x) = 1 = \sigma(x_0; x) \quad x \in [a, b]$$

in view of (4.30).

However,

$$|F_i - F_{i-1}| \leqslant \omega\left(f, x_{i-1}; \frac{2(b-a)}{m}\right) \leqslant \omega(f, x_{i-1}; 2\sqrt{d})$$

Therefore for every $x \in \Delta_{i_0}$ we have

$$\sum_{i=1}^{m} |F_i - F_{i-1}| |L(\sigma(x_{i-1}; t); x) - \sigma(x_{i-1}; x)|$$

$$\leqslant \sum_{i=2}^{m} \omega(f, x_{i-1}; 2\sqrt{d}) \alpha_i(x) \tag{4.46}$$

where

$$\alpha_i(x) = |L(\sigma(x_{i-1}; t); x) - \sigma(x_{i-1}; x)|$$

$$\alpha_i(x) \geqslant 0$$

$$\sum_{i=2}^{m} \alpha_i(x) \leqslant \sum_{i=2}^{i_0-1} \alpha_i(x) + 1 + 1 + \sum_{i=i_0+2}^{m} \alpha_i(x)$$

$$\leqslant \frac{4\pi^2}{3} + 2 \leqslant 16$$

in view of (4.43), (4.45) and (4.44).

Using Jensen's inequality (see 4.1), from (4.42) and (4.46) we obtain

$$|L(F; x) - F(x)|^p \leqslant \left| \sum_{i=2}^{m} \omega(f, x_{i-1}; 2\sqrt{d}) \alpha_i(x) \right|^p$$

$$\leqslant 14^{p-1} \sum_{i=2}^{m} \omega(f, x_{i-1}; 2\sqrt{d})^p \alpha_i(x) \tag{4.47}$$

From this, in view of Lemma 4.5 we obtain

$$\|LF - F\|_p \leqslant 14^{1-1/p} \left[\sum_{i=2}^{m} \omega(f, x_{i-1}; 2\sqrt{d})^p \int_a^b \alpha_i(x)\, dx \right]^{1/p}$$

$$\leqslant 14^{1-1/p} 4^{1/p} \left[\sum_{i=2}^{m} \omega(f, x_{i-1}; 2\sqrt{d})^p \sqrt{d} \right]^{1/p}$$

$$\leqslant 14 \left\{ \sum_{i=2}^{m} \int_{x_{i-1}}^{x_i} \omega(f, x_{i-1}; 2\sqrt{d})^p\, dx \frac{\sqrt{dm}}{b-a} \right\}^{1/p}$$

$$\leqslant 14 \times 2^{1/p} \left[\sum_{i=2}^{m} \int_{x_{i-1}}^{x_i} \omega(f, x; 4\sqrt{d})^p\, dx \right]^{1/p}$$

$$\leqslant 28 \left[\int_a^b \omega(f, x; 4\sqrt{d})^p\, dx \right]^{1/p}$$

$$= 28\tau(f; 4\sqrt{d})_p \leqslant 112\tau(f; \sqrt{d})_p \tag{4.48}$$

since $\omega(f, x_i; 2\sqrt{d}) \leqslant \omega(f, x; 4\sqrt{d})$, $x \in [x_{i-1}, x_i]$. We then obtain

$$\|LF - F\|_p \leqslant 112\tau(f; \sqrt{d})_p \qquad (4.49)$$

In the same way we find that

$$\|LG - G\|_p \leqslant 112\tau(f; \sqrt{d})_p \qquad (4.50)$$

The inequalities (4.40), (4.49) and (4.50) then give

$$\|Lf - f\|_{L_p[a,b]} \leqslant 368\tau(f; \sqrt{d})_{L_p[a,b]} \qquad \blacksquare$$

Notice that the method of the proof essentially increases the constant before $\tau(f; \sqrt{d})_p$.

Estimates for $\|Lf - f\|_p$, when L is a linear positive operator, are given by many authors, but in terms of the uniform moduli of smoothness of f or by the norms in $C[a, b]$ of derivatives of f. In our opinion the estimate from Theorem 4.3 by $\tau(f; \delta)_p$ is the natural estimate in L_p.

4.3. Approximation of functions by interpolating splines in L_p

In the last few years the interest in the approximation of functions by spline functions, which interpolate the approximating function at given points, has increased due to the following reasons:

(a) interpolating splines possess very good approximation properties;
(b) they are solutions of some classical extremal problems;
(c) the numerical evaluation of interpolating splines is easier and more convenient than the corresponding evaluation of interpolating polynomials.

In this section we shall give estimates for the approximation of functions by means of interpolating splines of first, second and third degree in L_p, $1 \leqslant p \leqslant \infty$.

It is very easy to obtain an estimate for the approximation of functions by means of interpolating splines of first degree, i.e. by means of polygons.

Let f be a bounded measurable function in the interval $[0, 1]$ and let $\{x_i\}_{i=0}^n$, $0 = x_0 < x_1 < \cdots < x_n = 1$ be $n + 1$ points in the interval $[0, 1]$. Let S_1 be the unique polygon which interpolates the function f at the points $\{x_i\}_{i=0}^n$, i.e. $f(x_i) = S_1(x_i)$ for $i = 0, 1, \ldots, n$, where S_1 is linear and continuous in $[x_{i-1}, x_i]$, $i = 1, \ldots, n$. The following estimate holds.

Theorem 4.4. *Let f be a bounded measurable function in the interval $[0, 1]$ and let S_1 be its interpolating spline of first degree with knots at the points $\{x_i\}_{i=0}^n$. Then*

$$\|f - S_1\|_{L_p[0,1]} \leqslant \tau_2(f; d_n)_{L_p[0,1]} \quad 1 \leqslant p \leqslant \infty$$

where

$$d_n = \max\{|x_i - x_{i-1}|: 1 \leqslant i \leqslant n\}$$

Proof. Since $f(x_{i-1}) = S_1(x_{i-1})$, $f(x_i) = S_1(x_i)$ and S_1 is a linear function in $[x_{i-1}, x_i]$, then, using Lemma 2.3, we have for every $x \in [x_{i-1}, x_i]$

$$|f(x) - S_1(x)| \leqslant \omega_2(f; [x_{i-1}, x_i])$$
$$\leqslant \omega_2(f, x; \Delta_i)$$
$$\leqslant \omega_2(f, x; d_n) \qquad (4.51)$$

where $\Delta_i = x_i - x_{i-1}$.

Taking the L_p-norm of both sides of (4.51), we obtain the statement of Theorem 4.4. ∎

We shall consider now interpolation by means of splines of higher degree. Let us remember that the function S, defined in the interval $[0, 1]$, is called a spline of kth degree with knots at the points $\{x_i\}_{i=0}^n$, $0 \leqslant x_0 < x_1 < \cdots < x_n = 1$, if $S \in C^{k-1}[0, 1]$ (i.e. S has continuous derivatives up to the order $k - 1$), and in each interval $[x_{i-1}, x_i]$, $i = 1, \ldots, n$, S is an algebraic polynomial of kth degree.

Let us first consider interpolation by means of a quadratic spline, i.e. the case $k = 2$. In this case it is convenient to choose as interpolation knots the points i/n, $i = 0, 1, \ldots, n$, and to choose as knots for the quadratic spline the points $(2i + 1)/2n$, $i = -3, -2, -1, 0, 1, \ldots, n, n+1, n+2, n+3$. It is well known (see Stećkin and Subbotin [1]) that every quadratic spline with knots at the points $(2i + 1)/2n$, $i = -3, -2, \ldots, n+2, n+3$ has the form

$$S_2(f; x) = \sum_{i=-1}^{n+1} c_i B_{2,i}(x) \qquad (4.52)$$

where

$$B_{2,i}(x) = B_{2,0}(x - ih) \quad h = 1/n$$

$$B_{2,0}(x) = \frac{n^2}{2}\left[\left(x + \frac{3}{2n}\right)_+^2 - 3\left(x + \frac{1}{2n}\right)_+^2 + 3\left(x - \frac{1}{2n}\right)_+^2 - \left(x - \frac{3}{2n}\right)_+^2\right]$$

$$(4.53)$$

are the so-called B-splines of Schoenberg.

Here, as usual, we set

$$(x)_+^k = \begin{cases} x^k & x \geqslant 0 \\ 0 & x < 0 \end{cases}$$

If is easy to verify that the B-spline (4.53) has the following properties (see Ahlberg et al. [1]):

$$B_{2,i}(x) \geqslant 0 \quad \text{for every } x, \ B_{2,i}(x) \leqslant \tfrac{3}{4}$$

$$\sum_{i=-1}^{n+1} B_{2,i}(x) = 1 \quad \text{for every } x \in [0, 1]$$

$$B_{2,i}(x) = 0 \quad \text{for } x \leqslant (i - \tfrac{3}{2})h \text{ and for } x \geqslant (i + \tfrac{3}{2})h \qquad (4.54)$$

$$B_{2,i}(ih) = \tfrac{3}{4}$$

$$B_{2,i}[(i-1)h] = B_{2,i}[(i+1)h] = \tfrac{1}{8}$$

$$\int_{-\infty}^{\infty} B_{2,i}(x)\,dx = \frac{1}{n}$$

Let f be a bounded measurable function in the interval $[0, 1]$. For simplicity we shall assume that the function f is 1-periodic (this is important only for the boundary conditions).

We require that the spline $S_2(f; x)$ should interpolate the function f at the points $x_i = i/n$, $i = 0, 1, \ldots, n$, i.e.

$$S_2(f; x_i) = f(x_i) \quad i = 0, 1, \ldots, n \tag{4.55}$$

and S_2 should satisfy the boundary conditions

$$\begin{aligned} S_2'(f; 0) &= S_2'(f; 1) \\ S_2''(f; 0) &= S_2''(f; 1) \end{aligned} \tag{4.56}$$

(These conditions come from the 1-periodicity of f.)

In view of the representation (4.52) of S_2 and the properties (4.54) of $B_{2,i}$, conditions (4.55) and (4.56) can be rewritten in the following way (see Stečkin and Subbotin [1]):

$$\tfrac{1}{8}c_{i-1} + \tfrac{3}{4}c_i + \tfrac{1}{8}c_{i+1} = f_i \quad f_i = f(x_i), i = 0, \ldots, n \tag{4.57}$$

$$\begin{aligned} c_{-1} &= c_0 + c_{n-1} - c_n \\ c_{n+1} &= -c_0 + c_1 + c_n \end{aligned} \tag{4.58}$$

Equations (4.57) and (4.58) are a system of linear algebraic equations for the unknowns c_i, $i = -1, 0, \ldots, n + 1$. If we exclude c_{-1} and c_{n+1} from (4.57) and (4.58), we obtain the system

$$\begin{aligned} c_{i-1} + 6c_i + c_{i+1} &= 8f_i \quad i = 1, \ldots, n-1 \\ 7c_0 + c_1 + c_{n-1} - c_n &= 8f_0 \\ -c_0 + c_1 + c_{n-1} + 7c_n &= 8f_n \end{aligned} \tag{4.59}$$

The system (4.59) is diagonally dominant. For such systems the following lemma holds:

Lemma 4.6. *Let*

$$x_i + \sum_{\substack{j=1 \\ j \neq i}}^{n} \varepsilon_{ij} x_j = f_i \quad i = 1, 2, \ldots, n$$

be a system of diagonally dominant linear algebraic equations. Let

$$\sum_{\substack{j=1 \\ j \neq i}}^{n} |\varepsilon_{ij}| \leqslant q < 1$$

$$\sum_{\substack{i=1 \\ i \neq j}}^{n} |\varepsilon_{ij}| \leqslant q < 1$$

Then

$$\|x\|_{l_p} \leqslant (1 - q)^{-1} \|f\|_{l_p} \tag{4.60}$$

where

$$\|x\|_{l_p} = \left[\sum_{i=1}^{n} |x_i|^p \right]^{1/p}$$

Proof. Let us write

$$q_i = \sum_{\substack{j=1 \\ j \neq i}}^{n} |\varepsilon_{ij}|$$

$$y_i = \sum_{\substack{j=1 \\ j \neq i}}^{n} \varepsilon_{ij} x_j$$

Then

$$\|x\|_{l_p} = \|f - y\|_{l_p} \leq \|f\|_{l_p} + \|y\|_{l_p} \qquad (4.61)$$

Using Jensen's inequality for $\|y\|_{l_p}$, we have

$$\|y\|_{l_p} = \left[\sum_{i=1}^{n} \left| \sum_{\substack{j=1 \\ j \neq i}}^{n} \varepsilon_{ij} x_j \right|^p \right]^{1/p}$$

$$\leq \left[\sum_{i=1}^{n} \left(\sum_{\substack{j=1 \\ j \neq i}}^{n} q_i \frac{|\varepsilon_{ij}| |x_j|}{q_i} \right)^p \right]^{1/p}$$

$$\leq \left[\sum_{i=1}^{n} \sum_{\substack{j=1 \\ j \neq i}}^{n} \frac{|\varepsilon_{ij}|}{q_j} |x_j|^p q_j^p \right]^{1/p}$$

$$= \left[\sum_{j=1}^{n} |x_j|^p q_j^{p-1} \sum_{\substack{i=1 \\ i \neq j}}^{n} |\varepsilon_{ij}| \right]^{1/p}$$

$$\leq \left[q^p \sum_{j=1}^{n} |x_j|^p \right]^{1/p} = q \|x\|_{l_p}$$

(4.60) follows from this inequality and (4.61). ∎

Let us now apply Lemma 4.6 to the system (4.59). This system can be rewritten in the form

$$\tfrac{1}{6} c_{i-1} + c_i + \tfrac{1}{6} c_{i+1} = \tfrac{8}{6} f_i \quad i = 1, 2, \ldots, n-1$$
$$c_0 + \tfrac{1}{7} c_1 + \tfrac{1}{7} c_{n-1} - \tfrac{1}{7} c_n = \tfrac{8}{7} f_0$$
$$\tfrac{1}{7}(-c_0 + c_1 + c_{n-1}) + c_n = \tfrac{8}{7} f_n$$

therefore it satisfies the conditions of the lemma with $q = 19/42$. Hence we have the estimate

$$\left[\sum_{i=0}^{n} |c_i|^p \right]^{1/p} \leq \frac{42}{23} \left[\sum_{i=0}^{n} (\tfrac{8}{6})^p |f_i|^p \right]^{1/p}$$

i.e.

$$\left[\sum_{i=0}^{n} |c_i|^p \right]^{1/p} \leq \frac{56}{23} \left[\sum_{i=0}^{n} |f_i|^p \right]^{1/p} \qquad (4.62)$$

Our aim will be to apply interpolation Theorem 2.9 from Chapter 2. First we shall estimate $\|S_2\|_p$ using (4.62). Also using (4.58) we get

$$\|S_2(f;\cdot)\|_{L_p[0,1]} = \left[\int_0^1 \left|\sum_{i=-1}^{n+1} c_i B_{2,i}(x)\right|^p dx\right]^{1/p}$$

$$= \left\{\int_0^1 \left| c_0[B_{2,-1}(x) + B_{2,0}(x) - B_{2,n+1}(x)]\right.\right.$$

$$+ c_1[B_{2,1}(x) + B_{2,n+1}(x)]$$

$$+ \sum_{i=2}^{n-2} c_i B_{2,i}(x) + c_{n-1}(B_{2,n-1}(x) + B_{2,-1}(x))$$

$$\left.\left. + c_n[B_{2,n}(x) + B_{2,-1}(x) + B_{2,n+1}(x)]\right|^p dx\right\}^{1/p}$$

$$\leqslant \left(\int_0^1 \left\{|c_0|[B_{2,-1}(x) + B_{2,0}(x) + B_{2,n+1}(x)]\right.\right.$$

$$+ |c_1|[B_{2,1}(x) + B_{2,n+1}(x)]$$

$$+ \sum_{i=2}^{n-2} |c_i| B_{2,i}(x) + |c_{n-1}|[B_{2,n-1}(x) + B_{2,-1}(x)]$$

$$\left.\left. + |c_n|[B_{2,n}(x) + B_{2,-1}(x) + B_{2,n+1}(x)]\right\}^p dx\right)^{1/p} \qquad (4.63)$$

Since $0 \leqslant B_{2,i}(x) \leqslant \frac{3}{4}$,

$$\sum_{i=-1}^{n+1} B_{2,i}(x) = 1$$

(see (4.54)), then for the sum of the coefficients of c_i, $i = 0, \ldots, n$, we have for $x \in [0,1]$

$$0 \leqslant B_{2,-1}(x) + B_{2,0}(x) + B_{2,n+1}(x) + B_{2,1}(x) + B_{2,n+1}(x)$$

$$+ \sum_{i=2}^{n-2} B_{2,i}(x) + B_{2,n-1}(x) + B_{2,-1}(x) + B_{2,n}(x) + B_{2,-1}(x) + B_{2,n+1}(x)$$

$$= \sum_{i=-1}^{n+1} B_{2,i}(x) + 2B_{2,n+1}(x) + 2B_{2,-1}(x)$$

$$\leqslant 1 + 2B_{2,n+1}(1) + 2B_{2,-1}(0) = 1 + \tfrac{1}{4} + \tfrac{1}{4} = \tfrac{3}{2}$$

Therefore, using Jensen's inequality, we obtain from (4.63), for $1 \leqslant p \leqslant \infty$,

$$\|S_2(f;\cdot)\|_p \leqslant \left(\int_0^1 (\tfrac{3}{2})^{p-1}\left\{|c_0|^p[B_{2,-1}(x) + B_{2,0}(x) + B_{2,n+1}(x)]\right.\right.$$

$$+ |c_1|^p[B_{2,1}(x) + B_{2,n+1}(x)] + \sum_{i=2}^{n-2} |c_i|^p B_{2,i}(x)$$

$$+ |c_{n-1}|^p [B_{2,n-1}(x) + B_{2,-1}(x)]$$

$$+ |c_n|^p [B_{2,n}(x) + B_{2,-1}(x) + B_{2,n+1}(x)] \Bigg\} dx \Bigg)^{1/p}$$

Since (see (4.53), (4.54))

$$\int_0^1 B_{2,i}(x)\, dx \leqslant \frac{1}{n} \quad i = 1, 2, \ldots, n-2, n-1$$

$$\int_0^1 B_{2,-1}(x)\, dx = \int_0^1 B_{2,n+1}(x)\, dx \leqslant \frac{1}{2n}$$

$$\int_0^1 B_{2,0}(x)\, dx = \int_0^1 B_{2,n}(x)\, dx \leqslant \frac{1}{2n}$$

we obtain

$$\| S_2(f;\cdot) \|_p \leqslant (\tfrac{3}{2})^{1-1/p} 2^{1/p} \left(\sum_{i=0}^n |c_i|^p n^{-1} \right)^{1/p}$$

$$\leqslant 2 \left(\sum_{i=0}^n |c_i|^p n^{-1} \right)^{1/p}$$

Finally we have

$$\| S_2(f;\cdot) \|_p \leqslant 2 \left(\sum_{i=0}^n |c_i|^p n^{-1} \right)^{1/p} \tag{4.64}$$

Inequalities (4.62) and (4.64) give us the following lemma.

Lemma 4.7. *For $1 \leqslant p \leqslant \infty$ we have*

$$\| S_2(f;\cdot) \|_p \leqslant 5 \left(\sum_{i=0}^n |f_i|^p n^{-1} \right)^{1/p}$$

This is the first inequality we need. Before we go further, let us give some consequences. In the case $p = \infty$ we obtain the following.

Corollary 4.4. *For the interpolating spline $S_2(f;\cdot)$ we have*

$$\| S_2(f;\cdot) \|_{C[0,1]} \leqslant 5 \max_{0 \leqslant i \leqslant n} |f_i|$$

or, since

$$f_i = S_2(f; x_i) \quad i = 0, 1, \ldots, n$$

$$\| S_2(f;\cdot) \|_{C[0,1]} \leqslant 5 \max_{0 \leqslant i \leqslant n} |S_2(f; x_i)| \tag{4.65}$$

This inequality is important for estimating the uniform norm of the spline with equidistant knots using the value of the spline at the mid-points between the knots.

For the case $1 \leqslant p < \infty$ we have the following.

Corollary 4.5.   *We have*

$$\|S_2(f;\cdot)\|_p \leqslant 5\left[\sum_{i=0}^{n}\left|S_2\left(f;\frac{i}{n}\right)\right|^p n^{-1}\right]^{1/p}$$

Theorem 4.5.   *Let the 1-periodic function f have p-integrable third derivative in the interval $[0,1]$, $1 \leqslant p \leqslant \infty$. Then*

$$\|S_2(f;\cdot) - f\|_{L_p[0,1]} \leqslant 10n^{-3}\|f'''\|_{L_p[0,1]}$$

This theorem is proved for the case $p = 2$ by Stečkin and Subbotin [1]. We shall give here the general proof.

Proof of Theorem 4.5.   From the periodic boundary conditions (4.56) for $S_2(f;\cdot)$ it follows that $S_2(f;\cdot)$ can also be considered as a 1-periodic function. The spline S_2 has a second derivative equal to a constant in every interval $((2i-1)/2n, (2i+1)/2n)$. Let $S_2''(f;x) = M_i$ for $x \in ((i-\frac{1}{2})/n, (i+\frac{1}{2})/n)$.

For the second finite difference, at the point x_{i-1} with step $h = 1/n$, of an arbitrary function g with $g'' \in L$, we have

$$\Delta^2 g_{i-1} = \int_0^h \int_0^h g''(x_{i-1} + t_1 + t_2)\,dt_1\,dt_2$$

Let us apply this equality for $g = S_2$. Since $S_2''(f;x) = M_{i-1}$ for $x \in ((i-1)/n, (i-\frac{1}{2})/n)$, $S_2''(f;x) = M_i$ for $x \in ((i-\frac{1}{2})/n, (i+\frac{1}{2})/n)$ and $S_2''(f;x) = M_{i+1}$ for $x \in ((i+\frac{1}{2})/n, (i+1)/n)$, we get

$$\Delta^2 S_2(f;x_{i-1})$$

$$= \int_0^h \int_0^h S_2''(f;x_{i-1} + t_1 + t_2)\,dt_1\,dt_2$$

$$= \iint_{\substack{0 \leqslant t_1+t_2 \leqslant h/2 \\ 0 \leqslant t_1, 0 \leqslant t_2}} S_2''(f;x_{i-1} + t_1 + t_2)\,dt_1\,dt_2$$

$$+ \iint_{\substack{h/2 \leqslant t_1+t_2 \leqslant 3h/2 \\ 0 \leqslant t_1 \leqslant h, 0 \leqslant t_2 \leqslant h}} S_2''(f;x_{i-1} + t_1 + t_2)\,dt_1\,dt_2$$

$$+ \iint_{\substack{3h/2 \leqslant t_1+t_2 \leqslant 2h \\ 0 \leqslant t_1 \leqslant h, 0 \leqslant t_2 \leqslant h}} S_2''(f;x_{i-1}, t_1 + t_2)\,dt_1\,dt_2$$

$$= \frac{h^2}{8}M_{i-1} + \frac{6h^2}{8}M_i + \frac{h^2}{8}M_{i+1}$$

Since $S_2(f;x_i) = f(x_i), i = -1,\ldots,n+1$, then

$$\Delta^2 S_2(f;x_{i-1}) = \Delta^2 f_{i-1}$$

and the previous equation shows that the M_i satisfies the system of linear algebraic equations:

$$\tfrac{1}{6}M_{i-1} + M_i + \tfrac{1}{6}M_{i+1} = \tfrac{8}{6}\Delta^2 f_{i-1}h^{-2} \quad i = 1,\ldots,n$$

(from the periodicity we have $M_0 = M_n$, $M_1 = M_{n+1}$).

Let us set

$$\varepsilon_i = M_i - f_i'' \quad f_i'' = f''\left(\frac{i}{n}\right)$$

Then the ε_i satisfy the following system of linear algebraic equations:

$$\tfrac{1}{6}\varepsilon_{i-1} + \varepsilon_i + \tfrac{1}{6}\varepsilon_{i+1} = \tfrac{1}{6}(8h^{-2}\Delta^2 f_{i-1} - f_{i-1}'' - 6f_i'' - f_{i+1}'') \tag{4.66}$$

where $i = 1,\ldots,n$, $\varepsilon_0 = \varepsilon_n$, $\varepsilon_1 = \varepsilon_{n+1}$.

The system (4.66) is diagonally dominant with $q = \tfrac{1}{3}$, and by Lemma 4.6 we have

$$\left(\sum_{i=1}^{n} |\varepsilon_i|^p\right)^{1/p} \leqslant \tfrac{3}{2}\left(\sum_{i=1}^{n} 6^{-p}|8h^{-2}\Delta^2 f_{i-1} - f_{i-1}'' - 6f_i'' - f_{i+1}''|^p\right)^{1/p}$$

But $h^{-2}\Delta^2 f_{i-1} = f''(\xi_i)$ for some $\xi_i \in ((i-1)/n, (i+1)/n)$ (f'' is continuous, since $f''' \in L_p[0,1]$) and therefore we have (with $1/p + 1/q = 1$)

$$\left(\sum_{i=1}^{n} |\varepsilon_i|^p\right)^{1/p} \leqslant \tfrac{3}{12}\left(\sum_{i=1}^{n} |8f''(\xi_i) - f_{i-1}'' - 6f_i'' - f_{i+1}''|^p\right)^{1/p}$$

$$= \tfrac{1}{4}\left(\sum_{i=1}^{n} \left|\int_{(i-1)/n}^{\xi_i} f'''(t)\,dt + 6\int_{i/n}^{\xi_i} f'''(t)\,dt + \int_{(i+1)/n}^{\xi_i} f'''(t)\,dt\right|^p\right)^{1/p}$$

$$\leqslant \tfrac{1}{4}\left\{\sum_{i=1}^{n}\left[8\int_{(i-1)/n}^{(i+1)/n} |f'''(t)|\,dt\right]^p\right\}^{1/p}$$

$$\leqslant 2\left[\sum_{i=1}^{n} (2n^{-1})^{p/q}\int_{(i-1)/n}^{(i+1)/n} |f'''(t)|^p\,dt\right]^{1/p}$$

$$\leqslant 2 \times 2^{1/p} \times 2^{1/q}n^{-1/q}\|f'''\|_p = 4n^{-1/q}\|f'''\|_p$$

Therefore

$$\sum_{i=1}^{n} |M_i - f_i''|^p \leqslant 4^p n^{-p/q}(\|f'''\|_{L_p[0,1]})^p \tag{4.67}$$

Let us set $\varphi = f - S_2$ and estimate $\|\varphi''\|_{L_p[0,1]}$. Since $S_2''(f;x) = M_i$ for $x \in ((i-\tfrac{1}{2})/n, (i+\tfrac{1}{2})/n)$, $i = 0,\ldots,n$, we obtain

$$\|\varphi''\|_p = \left[\int_0^1 |f''(x) - S_2''(x)|^p\,dx\right]^{1/p}$$

$$= \left[\int_{1/2n}^{1+1/2n} |f''(x) - S_2''(x)|^p\,dx\right]^{1/p}$$

$$= \left[\sum_{i=1}^{n}\int_{(i-1/2)/n}^{(i+1/2)/n} |f''(x) - M_i|^p\,dx\right]^{1/p} \tag{4.68}$$

Using the fact that $M_0 = M_n$ and (4.47), from (4.68) we obtain

$$\|\varphi''\|_p \leqslant 2^{1-1/p} \left[\sum_{i=1}^{n} \int_{(i-1/2)/n}^{(i+1/2)/n} |f''(x) - f_i''|^p \, dx + \sum_{i=1}^{n} \int_{(i-1/2)/n}^{(i+1/2)/n} |f_i'' - M_i|^p \, dx \right]^{1/p}$$

$$\leqslant 2^{1-1/p} \left[\sum_{i=1}^{n} \int_{(i-1/2)/n}^{(i+1/2)/n} \left| \int_{i/n}^{x} f'''(t) \, dt \right|^p dx + 4^p n^{-p/q} \|f'''\|_{L_p[0,1]}^p n^{-1} \right]^{1/p}$$

$$\leqslant 2^{1-1/p} \left[\sum_{i=1}^{n} n^{-1-p/q} \int_{(i-1/2)/n}^{(i+1/2)/n} |f'''(t)|^p \, dt + 4^p n^{-1-p/q} \|f'''\|_{L_p[0,1]}^p \right]^{1/p}$$

$$\leqslant 2^{1-1/p}(1+4) n^{-1/p-1/q} \|f'''\|_p$$

$$\leqslant 10 n^{-1} \|f'''\|_{L_p[0,1]}$$

This gives us the following theorem.

Theorem 4.6. *Let the 1-periodic function f have a p-integrable third derivative in $[0,1]$. Then*

$$\|f'' - S_2''(f;\cdot)\|_{L_p[0,1]} \leqslant 10 n^{-1} \|f'''\|_{L_p[0,1]}$$

Having this estimate for $\|\varphi''\|_p$, we can easily estimate φ'. The function φ is zero at the points i/n, $i = 0, 1, \ldots, n$, in view of the interpolation conditions. Hence its derivative has zeros at n points η_i, $\eta_i \in ((i-1)/n, i/n)$, $i = 1, \ldots, n$. If $x \in ((i-1)/n, i/n)$, then

$$\varphi'(x) = \int_{\eta_i}^{x} \varphi''(t) \, dt$$

and we have

$$\|\varphi'\|_p = \left[\int_0^1 |\varphi'(x)|^p \, dx \right]^{1/p}$$

$$= \left[\sum_{i=1}^{n} \int_{(i-1)/n}^{i/n} \left| \int_{\eta_i}^{x} \varphi''(t) \, dt \right|^p dx \right]^{1/p}$$

$$\leqslant \left[\sum_{i=1}^{n} \int_{(i-1)/n}^{i/n} n^{-p/q} \int_{(i-1)/n}^{i/n} |\varphi''(t)|^p \, dt \, dx \right]^{1/p} = n^{-1} \|\varphi''\|_p$$

We have thus obtained the following.

Theorem 4.7. *If the 1-periodic function f has a p-integrable third derivative in $[0,1]$, then*

$$\|f' - S_2'\|_{L_p[0,1]} \leqslant 10 n^{-2} \|f'''\|_{L_p[0,1]}$$

In the same way, since $\varphi(i/n) = 0$, $i = 0, 1, \ldots, n$, we have

$$\|f - S_2\|_p = \|\varphi\|_p$$

$$= \left[\sum_{i=1}^{n} \int_{(i-1)/n}^{i/n} \left| \int_{(i-1)/n}^{x} \varphi'(t)\,dt \right|^p dx \right]^{1/p}$$

$$\leqslant \left[\sum_{i=1}^{n} \int_{(i-1)/n}^{i/n} n^{-p/q} \int_{(i-1)/n}^{i/n} |\varphi'(t)|^p \, dt \, dx \right]^{1/p} = n^{-1} \|\varphi'\|_p$$

Theorem 4.5 follows from this inequality and Theorem 4.7.

From Lemma 4.7 and Theorem 4.5, using the interpolation Theorem 2.9 from Chapter 2, we obtain the following.

Theorem 4.8. *Let f be a bounded measurable 1-periodic function in the interval $[0, 1]$. If $S_2(f; \cdot)$ is its interpolating quadratic spline with knots at the points $(i - \frac{1}{2})/n, i = -1, 0, 1, \ldots, n+1, n+2$, which satisfies the interpolation conditions (4.55) and (4.56), then*

$$\|f - S_2(f; \cdot)\|_{L_p[0,1]} \leqslant c\tau_3(f; n^{-1})_{L_p[0,1]}$$

where c is an absolute constant.

Remark. Notice that Stečkin and Subbotin [1] give estimates for $\|f - S_2\|_C$ using $\omega(f; \delta)$ and for $\|f - S_2\|_{L_i} i = 1, 2$, with only $\|f'''\|_{L_i}$. Theorem 4.8 gives us all possible estimates.

Moreover, even in the uniform case, Theorem 4.8 gives us the estimate

$$\|f - S_2(f; \cdot)\|_C \leqslant c'\omega_3(f; n^{-1})$$

which is more general than the corresponding estimates given by Stečkin and Subbotin [1].

Using the properties of the modulus $\tau_3(f; \delta)_p$, we obtain the following corollaries from Theorem 4.8.

Corollary 4.6. *If the 1-periodic function f has first derivative $f' \in L_p[0, 1]$, then*

$$\|f - S_2(f; \cdot)\|_p = O[n^{-1}\omega_2(f'; n^{-1})_p]$$

Corollary 4.7. *If the 1-periodic function f has a k-derivative $f^{(k)}$ of bounded variation, $k = 0, 1, 2$, then*

$$\|f - S_2(f; \cdot)\|_{L_1[0,1]} = O(n^{-k-1}V_0^1 f^{(k)}) \quad k = 0, 1, 2$$

In an analogous way we can consider the interpolating splines with non-periodic boundary conditions. We shall restrict ourselves here to only a brief consideration of one of the possible cases, which we shall use later to estimate the error of collocation methods for solving the boundary value problem for ordinary differential equations of second order.

Let us consider the quadratic interpolating spline S_2, which, instead of the periodic boundary conditions, satisfies the conditions

$$S_2''(f; 0) = \alpha, \quad S_2''(f; 1) = \beta \tag{4.69}$$

Then the conditions for the coefficients c_i, $i = -1, 0, \ldots, n+1$, become (compare with (4.55) and (4.56))

$$c_{i-1} + 6c_i + c_{i+1} = 8f(x_i) \quad i = 0, 1, \ldots, n$$
$$c_{-1} - 2c_0 + c_1 = h^2\alpha$$
$$c_{n+1} - 2c_n + c_{n-1} = h^2\beta$$

In this system we substitute c_{-1} and c_{n+1} from the last two equations and obtain

$$c_0 = f(0) - \frac{h^2\alpha}{8}$$

$$\tfrac{1}{6}c_{i-1} + c_i + \tfrac{1}{6}c_{i-1} = \frac{8f(x_i)}{6} \quad i = 1, 2, \ldots, h-1$$

$$c_n = f(1) - \frac{h^2\beta}{8}$$

This system is diagonally dominant with $q = \frac{1}{3}$ and using Lemma 4.6 we obtain

$$\max_{0 \leqslant i \leqslant n} |c_i| \leqslant \tfrac{3}{2} \max \left\{ \max_{1 \leqslant i \leqslant n-1} \tfrac{4}{3}|f(x_i)|, |f(0)| + \frac{h^2|\alpha|}{8}, |f(1)| + \frac{h^2|\beta|}{8} \right\}$$

or

$$\max_{0 \leqslant i \leqslant n} |c_i| \leqslant 2 \max_{0 \leqslant i \leqslant n} |f(x_i)| + \frac{3h^2 \max\{|\alpha|, |\beta|\}}{16} \tag{4.70}$$

For the coefficient c_{-1} we have

$$|c_{-1}| = |2c_0 - c_1 + h^2\alpha|$$
$$= \left| 2f(0) - \frac{h^2\alpha}{4} - c_1 + \alpha h^2 \right|$$
$$= |2f(0) - c_1 + \tfrac{3}{4}h^2\alpha|$$
$$\leqslant 4 \max_{0 \leqslant i \leqslant n} |f(x_i)| + h^2 \max\{|\alpha|, |\beta|\} \tag{4.71}$$

In the same way we obtain

$$|c_{n+1}| \leqslant 4 \max_{0 \leqslant i \leqslant n} |f(x_i)| + h^2 \max\{|\alpha|, |\beta|\} \tag{4.72}$$

From (4.70)–(4.72) it follows that

$$\max_{-1 \leqslant i \leqslant n+1} |c_i| \leqslant 4 \max_{0 \leqslant i \leqslant n} |f(x_i)| + h^2 \max\{|\alpha|, |\beta|\} \tag{4.73}$$

Since

$$S_2(f; x) = \sum_{i=-1}^{n+1} c_i B_{2,0}(x - ih)$$

then

$$|S_2(f;x)| \leqslant \max_{-1\leqslant i\leqslant n+1} |c_i| \sum_{i=-1}^{n+1} B_{2,0}(x-ih)$$

In view of

$$\sum_{i=-1}^{n+1} B_{2,0}(x-ih) \leqslant 1$$

(see (4.54)) and (4.73), from the above inequality we obtain the following.

Theorem 4.9. *The interpolating spline S_2 with boundary conditions (4.69) satisfies*

$$\|S_2(f;\cdot)\|_{C[0,1]} \leqslant 4 \max_{0\leqslant i\leqslant n} |f(x_i)| + h^2 \max\{|\alpha|,|\beta|\}$$

We shall now estimate $\|f - S_2(f;\cdot)\|_{C[0,1]}$ with boundary conditions

$$S_2''(f;0) = f''(0), \quad S_2''(f;1) = f''(1) \tag{4.74}$$

Again the second derivative of S_2 is a constant in any of the intervals $((i-\frac{1}{2})/n,(i+\frac{1}{2})/n)$, $i=0,1,\ldots,n$. Let $S_2''(f;x) = M_i$, $x\in((i-\frac{1}{2})/n,(i+\frac{1}{2})/n)$. As in the previous case, in view of the boundary conditions (4.74), we obtain the following equations for M_i:

$$\tfrac{1}{6}M_{i-1} + M_i + \tfrac{1}{6}M_{i+1} = \tfrac{8}{6}\Delta^2 f_{i-1}h^{-2} \quad i=1,\ldots,n-1$$
$$M_0 = f''(0), \quad M_n = f''(1)$$

Then for $\varepsilon_i = M_i - f_i''$, $f_i'' = f''(i/n)$, we have the following system:

$$\tfrac{1}{6}\varepsilon_{i-1} + \varepsilon_i + \tfrac{1}{6}\varepsilon_{i+1} = \tfrac{1}{6}(8h^{-2}\Delta^2 f_{i-1} - f_{i-1}'' - 6f_i'' - f_{i+1}'')$$

where $i=1,2,\ldots,n$, $\varepsilon_0 = \varepsilon_n = 0$.

This system is diagonally dominant with $q = \frac{1}{3}$, and therefore, in view of Lemma 4.6, we obtain

$$\max_{0\leqslant i\leqslant n} |\varepsilon_i| \leqslant \tfrac{3}{2} \max_{1\leqslant i\leqslant n-1} |\tfrac{8}{6}h^{-2}\Delta^2 f_{i-1} - \tfrac{1}{6}f_{i-1}'' - f_i'' - \tfrac{1}{6}f_{i+1}''|$$

$$= \tfrac{3}{2} \max_{1\leqslant i\leqslant n-1} |\tfrac{8}{6}f''(\xi_i) - \tfrac{1}{6}f_{i-1}'' - f_i'' - \tfrac{1}{6}f_{i+1}''|$$

$$\leqslant 2\omega\left(f'';\frac{2}{n}\right)$$

$$\leqslant 4\omega(f'';n^{-1}) \quad \xi_i\in\left(\frac{i-1}{n},\frac{i+1}{n}\right)$$

Consequently for $x\in((i-\frac{1}{2})/n,(i+\frac{1}{2})/n)$ we have

$$|f''(x) - S_2''(f;x)| = |f''(x) - M_i|$$
$$\leqslant |f''(x) - f_i''| + |f_i'' - M_i|$$
$$\leqslant \omega(f'';n^{-1}) + 4\omega(f'';n^{-1}) = 5\omega(f'';n^{-1})$$

i.e.

$$\sup_{x \in [0,1]} |f''(x) - S_2''(f; x)| \leqslant 5\omega(f''; n^{-1})$$

From the inequality, as in the proof of Theorems 4.7 and 4.8, we obtain successively

$$\|f' - S_2'\|_{C[0,1]} \leqslant 5n^{-1}\omega(f''; n^{-1})$$
$$\|f - S_2\|_{C[0,1]} \leqslant 5n^{-2}\omega(f''; n^{-1})$$

from which we get the following theorem.

Theorem 4.10. *Let S_2 be a quadratic interpolating spline for the function f, satisfying the boundary conditions $S_2''(f; 0) = f''(0)$, $S_2''(f; 1) = f''(1)$. Then*

$$\|f - S_2\|_{C[0,1]} \leqslant 5n^{-2}\omega(f''; n^{-1})$$

Now let us consider the interpolation of functions by means of cubic splines. In this case we can take for the knots of the cubic spline and for the interpolation knots the points $x_i = i/n$, $i = -3, -2, \ldots, n+2$. Every cubic spline S_3 in the interval $[0,1]$ with knots at the points x_i, $i = 0, 1, \ldots, n$, can be represented in the form (see Steckin and Subbotin [1])

$$S_3(f; x) = \sum_{i=-1}^{n+1} \alpha_i B_{3,i}(x) \tag{4.75}$$

where

$$B_{3,i}(x) = B_{3,0}(x - ih) \quad h = \frac{1}{n}$$

$$B_{3,0}(x) = \frac{n^3}{6} \sum_{k=0}^{4} (-1)^k \binom{4}{k} [x + (2 - k)h]_+^3 \tag{4.76}$$

is the Schoenberg B-spline of third degree.

It is easy to verify that the B-spline (4.76) has the properties

$$0 \leqslant B_{3,0}(x) \leqslant 1 \quad \text{for every } x$$

$$\int_{-\infty}^{\infty} B_{3,0}(x) \, dx = \int_{-\infty}^{\infty} B_{3,i}(x) \, dx = \frac{1}{n}$$

$$0 \leqslant \sum_{i=-1}^{n+1} B_{3,i}(x) \leqslant 1 \quad \text{for every } x \tag{4.77}$$

$$B_{3,i}(x) = 0 \quad \text{for } x \leqslant (i-2)h \text{ and for } x \geqslant (i+2)h$$
$$B_{3,i}(ih) = \tfrac{4}{6}, \quad B_{3,i}[(i-1)h] = B_{3,i}[(i+1)h] = \tfrac{1}{6}$$

Let f be a bounded measurable function in the interval $[0,1]$. We require that the spline S_3 should interpolate the function f at the points $x_i = i/n$, $i = 0, 1, \ldots, n$, i.e. it should satisfy the conditions

$$S_3(f; x_i) = f(x_i) = f_i \quad i = 0, 1, \ldots, n \tag{4.78}$$

We need also to choose boundary conditions. We shall consider boundary conditions of two types:

(a) periodic, in the case when the function f is 1-periodic:

$$S_3'(f;0) = S_3'(f;1), \quad S_3''(f;0) = S_3''(f;1) \tag{4.79}$$

(b) conditions on the second derivative of f at the two ends:

$$S_3''(f;0) = \alpha, \quad S_3''(f;1) = \beta \tag{4.80}$$

Usually such conditions are in posed when the function has a second derivative. Then we usually set

$$\alpha = f''(0), \quad \beta = f''(1)$$

In view of the representations (4.75), (4.76) and properties (4.77), from the interpolation conditions (4.78) and the boundary conditions (4.79) (respectively (4.80)), we obtain the following systems of linear algebraic equations for determining the coefficients α_i, $i = -1, 0, \ldots, n + 1$, in the representation (4.75):

(a) in the case of boundary conditions (4.79):

$$\alpha_{i-1} + 4\alpha_i + \alpha_{i+1} = 6f_i \quad i = 0, \ldots, n$$

$$\alpha_{-1} = \alpha_0 + \alpha_{n-1} - \alpha_n$$

$$\alpha_{n+1} = -\alpha_0 + \alpha_1 + \alpha_n$$

(b) in the case of boundary conditions (4.80):

$$\alpha_{i-1} + 4\alpha_i + \alpha_{i+1} = 6f_i$$

$$\alpha_{-1} - 2\alpha_0 + \alpha_1 = \alpha h^2$$

$$\alpha_{n-1} - 2\alpha_n + \alpha_{n+1} = \beta h^2$$

If we exclude α_{-1} and α_{n+1}, we obtain the system

(a)
$$\tfrac{1}{4}\alpha_{i-1} + \alpha_i + \tfrac{1}{4}\alpha_{i+1} = \tfrac{6}{4}f_i \quad i = 1, 2, \ldots, n-1$$

$$\alpha_0 + \tfrac{1}{5}\alpha_1 + \tfrac{1}{5}\alpha_{n-1} - \tfrac{1}{5}\alpha_n = \tfrac{6}{5}f_0 \tag{4.81}$$

$$-\tfrac{1}{5}\alpha_0 + \tfrac{1}{5}\alpha_1 + \tfrac{1}{5}\alpha_{n-1} + \alpha_n = \tfrac{6}{5}f_n$$

(b)
$$\tfrac{1}{4}\alpha_{i-1} + \alpha_i + \tfrac{1}{4}\alpha_{i+1} = \tfrac{6}{4}f_i \quad i = 1, \ldots, n-1$$

$$\alpha_0 = f_0 - \frac{\alpha h^2}{6} \tag{4.82}$$

$$\alpha_n = f_n - \frac{\beta h^2}{6}$$

These systems are diagonally dominant, with $q = \tfrac{3}{5}$ for the system (4.81) and $q = \tfrac{1}{2}$ for the system (4.82). If we apply Lemma 4.6, we obtain, in both cases,

(a)
$$\left(\sum_{i=0}^{n}|\alpha_i|^p\right)^{1/p} \leqslant \tfrac{5}{2}\left[\sum_{i=0}^{n}(\tfrac{6}{4})^p|f_i|^p\right]^{1/p}$$

$$= \tfrac{15}{4}\left(\sum_{i=0}^{n}|f_i|^p\right)^{1/p} \quad 1 \leqslant p \leqslant \infty$$

(b)
$$\left(\sum_{i=0}^{n}|\alpha_i|^p\right)^{1/p} \leqslant 2\left[\sum_{i=1}^{n-1}(\tfrac{6}{4})^p|f_i|^p + |f_0|^p + |f_n|^p\right]^{1/p}$$

$$+ 2\left[(|\alpha|^p + |\beta|^p)\frac{h^{2p}}{6^p}\right]^{1/p}$$

$$\leqslant 4\left(\sum_{i=0}^{n}|f_i|^p\right)^{1/p} + (|\alpha| + |\beta|)\frac{h^2}{3}$$

From these two inequalities, using representation (4.75), as in the case of quadratic splines and Jensen's inequality, we obtain (compare with (4.63) where we have $\sum_{i=-1}^{n+1} B_{3,i}(x) + 2B_{3,-1}(x) + B_{3,n+1}(x) \leqslant \tfrac{5}{3}$)

(a)
$$\|S_3(f;\cdot)\|_p \leqslant \tfrac{5}{3}3^{1/p}\left(\sum_{i=0}^{n}|\alpha_i|^p n^{-1}\right)^{1/p}$$

$$\leqslant 5\left(\sum_{i=0}^{n}|\alpha_i|^p n^{-1}\right)^{1/p}$$

$$\leqslant \tfrac{75}{4}\left(\sum_{i=0}^{n}|f_i|^p n^{-1}\right)^{1/p}$$

(b)
$$\|S_3(f;\cdot)\|_p \leqslant \tfrac{5}{3} \times 3^{1/p}\left(\sum_{i=0}^{n}|\alpha_i|^p n^{-1}\right)^{1/p}$$

$$\leqslant 20\left(\sum_{i=0}^{n}|f_i|^p n^{-1}\right)^{1/p} + 10(|\alpha| + |\beta|)\frac{h^2}{6}$$

We have thus obtained the following lemma.

Lemma 4.8. *For the interpolating cubic spline $S_3(f;\cdot)$ for the function f we have* $(1 \leqslant p \leqslant \infty)$

(a) *in the case of boundary conditions (4.79)*

$$\|S_3(f;\cdot)\|_p \leqslant \tfrac{75}{4}\left(\sum_{i=0}^{n}|f_i|^p n^{-1}\right)^{1/p}$$

(b) *in the case of the boundary conditions (4.80)*

$$\|S_3(f;\cdot)\|_p \leqslant 20\left(\sum_{i=0}^{n}|f_i|^p n^{-1}\right)^{1/p} + 10(|\alpha| + |\beta|)\frac{h^2}{6}$$

From Lemma 4.3 we can again obtain, as in the case of quadratic interpolating splines, some important corollaries.

Corollary 4.8. *For the interpolating cubic spline of the function f we have:*

(a) *in the case of boundary conditions (4.79)*

$$\|S_3(f;\cdot)\|_{C[0,1]} \leqslant \tfrac{75}{4} \max_{0 \leqslant i \leqslant n}|f_i|$$

(b) *in the case of boundary conditions* (4.80)

$$\|S_3(f;\cdot)\|_{C[0,1]} \leqslant 20 \max_{0 \leqslant i \leqslant n} |f_i| + \tfrac{5}{3}(|\alpha| + |\beta|)n^{-2}$$

Since $S_3(f;x) = f(x_i)$, $i = 0, 1, \ldots, n$, we can rewrite the above two inequalities in the form

$$\|S_3(f;\cdot)\|_{C[0,1]} \leqslant \tfrac{75}{4} \max_{0 \leqslant i \leqslant n} |S_3(f;x_i)|$$

$$\|S_3(f;\cdot)\|_{C[0,1]} \leqslant 20 \max_{0 \leqslant i \leqslant n} |S_3(f;x_i)| + \tfrac{5}{3}(|\alpha| + |\beta|)n^{-2}$$

Corollary 4.9. *In the case of the boundary conditions* (4.79) *we have*

$$\|S_3(f;\cdot)\|_{L_p[0,1]} \leqslant \tfrac{75}{4} \left[\sum_{i=0}^{n} |S_3(f;x_i)|^p n^{-1} \right]^{1/p}$$

In order to apply interpolation Theorem 2.9 we need the following theorem.

Theorem 4.11. *Let the function f have a fourth derivative f^{iv} in the interval* $[0, 1]$ *and $f^{iv} \in L_p[0, 1]$. Let $S_3(f;\cdot)$ be the interpolating cubic spline for f. Then*

(a) *if the function f is 1-periodic and the interpolating cubic spline satisfies the boundary conditions* (4.79), *then*

$$\|f''' - S_3'''\|_{L_p[0,1]} \leqslant 17n^{-1} \|f^{iv}\|_{L_p[0,1]}$$

(b) *if the interpolating cubic spline satisfies the boundary conditions* (4.80), *then*

$$\|f''' - S_3'''\|_{L_p[0,1]} \leqslant 17n^{-1} \|f^{iv}\|_{L_p[0,1]} + 2n^{1-1/p}[|\alpha - f''(0)| + |\beta - f''(1)|]$$

Proof. Using the notation $S_3''(f;x_i) = M_i$, $i = 0, 1, \ldots, n$, the M_i satisfy (see Stečkin and Subbotin [1])

$$M_{i-1} + 4M_i + M_{i+1} = 6h^{-2}\Delta^2 f_{i-1} \quad i = 1, 2, \ldots, n-1 \tag{4.83}$$

The two types of boundary conditions give us the following additional conditions::

(a) in the case of periodic boundary conditions (4.79)

$$M_0 = M_n, \quad M_1 = M_{n+1}$$

and (4.38) holds also for $i = n$;

(b) in the case of boundary conditions (4.80)

$$M_0 = \alpha, \quad M_n = \beta$$

Let us set:

$$\varepsilon_i = M_i - f_i'' \quad f_i'' = f''\left(\frac{i}{n}\right)$$

Then for ε_i we obtain the following systems of linear algebraic equations for the above cases respectively:

(a) $\qquad \frac{1}{4}\varepsilon_{i-1} + \varepsilon_i + \frac{1}{4}\varepsilon_{i+1} = \frac{1}{4}(6h^{-2}\Delta^2 f_{i-1} - f''_{i-1} - 4f''_i - f''_{i+1})$ (4.84)

where $i = 1, 2, \ldots, n - 1$, $\varepsilon_0 = \varepsilon_n$, $\varepsilon_1 = \varepsilon_{n+1}$;

(b) $\qquad \frac{1}{4}\varepsilon_{i-1} + \varepsilon_i + \frac{1}{4}\varepsilon_{i+1} = \frac{1}{4}(6h^{-2}\Delta^2 f_{i-1} - f''_{i-1} - 4f''_i - f''_{i+1})$, (4.85)

where $i = 1, 2, \ldots, n - 1$, $\varepsilon_0 = \alpha - f''(0)$, $\varepsilon_n = \beta - f''(1)$.

Let us estimate $g_i = 6h^{-2}\Delta^2 f_{i-1} - f''_{i-1} - 4f''_i - f''_{i+1}$ under the assumption that the function f has a fourth derivative $f^{iv} \in L_p[0, 1]$. We have

$$g_i = 6(h^{-2}\Delta^2 f_{i-1} - f''_i) - (f''_{i+1} - 2f''_i + f''_{i+1})$$

From Taylor's formula with integral remainder over obtain

$$h^{-2}\Delta^2 f_{i-1} - f''_i = \frac{h^{-2}}{3!} \int_{x_{i-1}}^{x_{i+1}} K_i(t) f^{iv}(t) \, dt$$

where

$$K_i(t) = \begin{cases} (x_{i+1} - t)^3 & t \in [x_i, x_{i+1}] \\ (t - x_{i-1})^3 & t \in [x_{i-1}, x_i] \end{cases}$$

Hence, applying Holder's inequality, we obtain (with $1/p + 1/q = 1$)

$$|h^{-2}\Delta^2 f_{i-1} - f''_i| \leqslant \frac{h^{(1+1/q)}}{6} \left[\int_{x_{i-1}}^{x_i} |f^{iv}(t)|^p \, dt \right]^{1/p}$$ (4.86)

On the other hand, again applying Taylor's formula with integral remainder, we obtain

$$f''_{i+1} - 2f''_i + f''_{i-1} = \Delta^2 f''_{i-1} = \int_{x_{i-1}}^{x_{i+1}} Q_i(t) f^{iv}(t) \, dt$$

where

$$Q_i(t) = \begin{cases} x_{i+1} - t & t \in [x_i, x_{i+1}] \\ t - x_{i-1} & t \in [x_{i-1}, x_i] \end{cases}$$

Hence

$$|\Delta^2 f''_{i-1}| \leqslant h^{1+1/q} \left[\int_{x_{i-1}}^{x_{i+1}} |f^{iv}(t)|^p \, dt \right]^{1/p}$$ (4.87)

Then the estimates (4.86) and (4.87) yield

$$|g_i| \leqslant 2h^{1+1/q} \left[\int_{x_{i-1}}^{x_{i+1}} |f^{iv}(t)| p \, dt \right]^{1/p}$$ (4.88)

Since the systems (4.84) and (4.85) have a dominating major diagonal with $q = \frac{1}{2}$, Lemma 4.6 together with (4.88) gives us:

(a) for the system (4.84)

$$\left(\sum_{i=1}^{n}|\varepsilon_i|^p\right)^{1/p}\leqslant 2\left(\sum_{i=1}^{n}|g_i|^p\right)^{1/p}$$

$$\leqslant 4h^{1+1/q}\left[\sum_{i=1}^{n}\int_{x_{i-1}}^{x_{i+1}}|f^{iv}(x)|^p\,dx\right]^{1/p}\leqslant 8h^{1+1/q}\|f^{iv}\|_p$$

i.e.

$$\left(\sum_{i=1}^{n}|\varepsilon_i|^p\right)^{1/p}\leqslant 8h^{1+1/q}\|f^{iv}\|_p \qquad (4.89)$$

(b) for the system (4.85) we have an additional term $|\alpha-f''(0)|+|\beta-f''(1)|$, i.e. the estimate is

$$\left(\sum_{i=1}^{n}|\varepsilon_i|^p\right)^{1/p}\leqslant 8h^{1+1/q}\|f^{iv}\|_p+|\alpha-f''(0)|+|\beta-f''(1)| \qquad (4.90)$$

Let us now estimate $\|\varphi'''\|_p$, where $\varphi=f-S_3$. We set $\sigma_i=[M_{i+1}-M_i]/h$. Then for $x\in(x_i,x_{i+1})$ we have $S_3'''(f;x)=\sigma_i$. Let us consider first the periodic case (a). We define the functions ψ_1,ψ_2,ψ_3 in the following way:

$$\psi_1(x)=\frac{\varepsilon_{i+1}}{h} \qquad\qquad \text{for } x\in[x_i,x_{i+1})$$

$$\psi_2(x)=\frac{\varepsilon_i}{h} \qquad\qquad \text{for } x\in[x_i,x_{i+1})$$

$$\psi_3(x)=\frac{f''_{i+1}-f''_i}{h}-f'''(x) \quad \text{for } x\in[x_i,x_{i+1})$$

Since $f'''-S_3'''=\psi_2-\psi_1-\psi_3$, then

$$\|\varphi'''\|_p\leqslant\|\psi_1\|_p+\|\psi_2\|_p+\|\psi_3\|_p \qquad (4.91)$$

We have

$$\|\psi_1\|_p=\left[\sum_{i=1}^{n}\int_{x_{i-1}}^{x_i}|\psi_1(x)|^p\,dx\right]^{1/p}$$

$$=\left(\sum_{i=1}^{n}|\varepsilon_i|^p h^{1-p}\right)^{1/p}$$

$$=h^{1/p-1}\left(\sum_{i=1}^{n}|\varepsilon_i|^p\right)^{1/p}$$

$$\|\psi_2\|_p=\left[\sum_{i=1}^{n}\int_{x_{i-1}}^{x_i}|\psi_2(x)|^p\,dx\right]^{1/p}$$

$$=\left(\sum_{i=1}^{n}|\varepsilon_{i-1}|^p h^{1-p}\right)^{1/p}$$

$$= h^{1/p-1}\left(\sum_{i=0}^{n-1}|\varepsilon_i|^p\right)^{1/p} \tag{4.92}$$

$$= h^{1/p-1}\left(\sum_{i=1}^{n}|\varepsilon_i|^p\right)^{1/p}$$

since $\varepsilon_0 = \varepsilon_n$;

$$\|\psi_3\|_p = \left[\sum_{i=1}^{n}\int_{x_{i-1}}^{x_i}\left|\frac{f_i'' - f_{i-1}''}{h} - f'''(x)\right|^p dx\right]^{1/p}$$

$$= \left[\sum_{i=1}^{n}\int_{x_{i-1}}^{x_i}\left|\int_{x}^{\xi_i}f^{iv}(t)\,dt\right|^p dx\right]^{1/p}$$

$$= \left\{\sum_{i=1}^{n}\int_{x_{i-1}}^{x_i}\left[\int_{x_{i-1}}^{x_i}|f^{iv}(t)|\,dt\right]^p dx\right\}^{1/p}$$

$$= h^{1/p}h^{1/q}\left[\sum_{i=1}^{n}\int_{x_{i-1}}^{x_i}|f^{iv}(t)|^p\,dt\right]^{1/p} = h\|f^{iv}\|_p$$

The estimates (4.92) together with (4.91) and (4.89) give

$$\|\varphi'''\|_p \leqslant 2h^{1/p-1}\left(\sum_{i=1}^{n}|\varepsilon_i|^p\right)^{1/p} + h\|f^{iv}\|_p$$

$$\leqslant 16h^{1/p+1/q}\|f^{iv}\|_p + h\|f^{iv}\|_p = 17n^{-1}\|f^{iv}\|_p$$

which is statement (a) of Theorem 4.11.

In the non-periodic case the estimates for ψ_i are the same:

$$\|\psi_1\|_p \leqslant h^{1/p-1}\left(\sum_{i=1}^{n}|\varepsilon_i|^p\right)^{1/p}$$

$$\|\psi_2\|_p \leqslant h^{1/p-1}\left(\sum_{i=0}^{n-1}|\varepsilon_i|^p\right)^{1/p}$$

$$\|\psi_3\|_p \leqslant h\|f^{iv}\|_p$$

which together with (4.91) and (4.90) give

$$\|\varphi'''\|_p \leqslant 16h^{1/p+1/q}\|f^{iv}\|_p + h\|f^{iv}\|_p$$

$$+ 2[|\alpha - f''(0)| + |\beta - f''(1)|]h^{1/p-1}$$

$$= 17h\|f^{iv}\|_p + 2h^{1/p-1}[|\alpha - f''(0)| + |\beta - f''(1)|] \qquad\blacksquare$$

Theorem 4.12. *Let the function f have a fourth derivative f^{iv} in the interval $[0, 1]$ and $f^{iv} \in L_p[0, 1]$. Let $S_3(f; x)$ be the interpolating cubic spline for the function f. Then for $1 \leqslant p \leqslant \infty$ we have*

(a) *if the function f is 1-periodic and the interpolating cubic spline satisfies the boundary conditions (4.79), then*

$$\|f'' - S_3''\|_{L_p[0,1]} \leqslant 18n^{-2}\|f^{iv}\|_{L_p[0,1]}$$

(b) *if the interpolating cubic spline satisfies the boundary conditions* (4.80), *then*

$$\| f'' - S_3'' \|_{L_p[0,1]} \leqslant 18n^{-2} \| f^{iv} \|_p + 2n^{-1/p}[|\alpha - f''(0)| + |\beta - f''(1)|]$$

Proof. We shall use the same notation as in the proof of Theorem 4.11. For $x \in [x_i, x_{i+1}]$ we have

$$f''(x) - S_3''(f; x) = f''(x) - f''(x_i)\frac{x_{i+1} - x}{h}$$

$$- f(x_{i+1})\frac{x - x_i}{h} - (M_i - f_i'')\frac{x_{i+1} - x}{h}$$

$$- (M_{i+1} - f_{i+1}'')\frac{x - x_i}{h} \tag{4.93}$$

Let us introduce the functions ψ_i, $i = 1, 2, 3$, defined by

$$\psi_1(x) = f''(x) - f''(x_i)\frac{x_{i+1} - x}{h} - f''(x_{i+1})\frac{x - x_i}{h} \quad \text{for } x \in [x_i, x_{i+1}]$$

$$\psi_2(x) = (M_i - f_i'')\frac{x_{i+1} - x}{h} \quad\quad\quad\quad\quad \text{for } x \in [x_i, x_{i+1}]$$

$$\psi_3(x) = (M_{i+1} - f_{i+1}'')\frac{x - x_i}{h} \quad\quad\quad\quad \text{for } x \in [x_i, x_{i+1}]$$

From (4.93) it follows that $f'' - S_3'' = \psi_1 - \psi_2 - \psi_3$.
Hence

$$\| S_3'' - f'' \|_p \leqslant \| \psi_1 \|_p + \| \psi_2 \|_p + \| \psi_3 \|_p \tag{4.94}$$

Let us estimate the norms $\| \psi_i \|_p$, $i = 1, 2, 3$. For $x \in [x_i, x_{i+1}]$, using Theorem 2.1, we have

$$|\psi_1(x)| \leqslant \omega_2\left(f'', \frac{x_i + x_{i+1}}{2}; h\right) \leqslant \omega_2(f'', x; h)$$

and therefore

$$\| \psi_1 \|_p \leqslant \tau_2(f''; h)_p$$

Using the properties of τ_2 from Chapter 1 (property (4)) we obtain

$$\| \psi_1 \|_p \leqslant 2h^2 \| f^{iv} \|_p \tag{4.95}$$

Using the definition of functions ψ_2 and ψ_3 we obtain

$$\| \psi_2 \|_p = \left[\sum_{i=0}^{n-1} \int_{x_i}^{x_{i+1}} |\psi_2(x)|^p \, dx \right]^{1/p}$$

$$\leqslant \left(\sum_{i=1}^{n-1} |\varepsilon_i|^p h \right)^{1/p}$$

$$= h^{1/p} \left(\sum_{i=0}^{n-1} |\varepsilon_i|^p \right)^{1/p}$$

$$\|\psi_3\|_p = \left[\sum_{i=0}^{n-1}\int_{x_i}^{x_{i+1}}|\psi_3(x)|^p\,dx\right]^{1/p}$$

$$\leqslant\left(\sum_{i=0}^{n-1}|\varepsilon_{i+1}|^p h\right)^{1/p}$$

$$= h^{1/p}\left(\sum_{i=1}^{n}|\varepsilon_i|^p\right)^{1/p}$$

Using the inequalities (4.89), (4.90) for the periodic an non-periodic cases respectively, we obtain the following.

In case (a)

$$\|\psi_2\|_p \leqslant 8h^{1/p+1/q+1}\|f^{iv}\|_p = 8n^{-2}\|f^{iv}\|_p$$

$$\|\psi_3\|_p \leqslant 8h^{1/p+1/q+1}\|f^{iv}\|_p = 8n^{-2}\|f^{iv}\|_p \tag{4.96}$$

In case (b)

$$\|\psi_2\|_p \leqslant 8h^{1/p+1/q+1}\|f^{iv}\|_p + h^{1/p}[|\alpha-f''(0)|+|\beta-f''(1)|]$$

$$= 8n^{-2}\|f^{iv}\|_p + n^{-1/p}[|\alpha-f''(0)|+|\beta-f''(1)|] \tag{4.97}$$

$$\|\psi_3\|_p \leqslant 8n^{-2}\|f^{iv}\|_p + n^{-1/p}[|\alpha-f''(0)|+|\beta-f''(1)|]$$

Inequalities (4.94), (4.95) and (4.96) give, in case (a),

$$\|f''-S_3''\|_p \leqslant 2n^{-2}\|f^{iv}\|_p + 16n^{-2}\|f^{iv}\|_p = 18n^{-2}\|f^{iv}\|_p$$

Inequalities (4.94), (4.95) and (4.97) give, in case (b),

$$\|f''-S_3''\|_p \leqslant 2n^{-2}\|f^{iv}\|_p + 16n^{-2}\|f^{iv}\|_p + 2n^{-1/p}[|\alpha-f''(0)|+|\beta-f''(1)|]$$

$$= 18n^{-2}\|f^{iv}\|_p + 2n^{-1/p}\{|\alpha-f''(0)|+|\beta-f''(1)|\}$$

Thus Theorem 4.12 is proved. ∎

Let us denote $\varphi = f - S_3$. As in the case of interpolating quadratic splines, using only the interpolation conditions $\varphi(i/n) = 0$, $i = 1,\ldots,n$, we obtain

$$\|\varphi'\|_p \leqslant n^{-1}\|\varphi''\|_p$$

$$\|\varphi\|_p \leqslant n^{-2}\|\varphi''\|_p$$

From these inequalities and Theorem 4.12 we obtain the following.

Theorem 4.13. *Let the function f have a fourth derivative f^{iv} in the interval $[0,1]$ and $f^{iv}\in L_p[0,1]$. Let $S_3(f;x)$ be the interpolating cubic spline for the function f with interpolation knots i/n, $i = 0, 1,\ldots,n$. Then for $1\leqslant p\leqslant\infty$ we have:*

(a) *if the function f is 1-periodic and the interpolating spline satisfies the periodic boundary conditions (4.79), then*

$$\|f-S_3\|_p \leqslant 18n^{-4}\|f^{iv}\|_p$$

$$\|f'-S_3'\|_p \leqslant 18n^{-3}\|f^{iv}\|_p$$

(b) *if the interpolating cubic spline satisfies the boundary conditions* (4.80), *then*

$$\|f - S_3\|_p \leqslant 18n^{-4}\|f^{iv}\|_p + 2n^{-2-1/p}[|\alpha - f''(0)| + |\beta - f''(1)|]$$

$$\|f' - S_3'\|_p \leqslant 18n^{-3}\|f^{iv}\|_p + 2n^{-1-1/p}[|\alpha - f''(0)| + |\beta - f''(1)|]$$

From Theorem 4.13, case (a), Lemma 4.8, case (a), and interpolation Theorem 2.9 we obtain the following theorem.

Theorem 4.14. *Let f be a bounded measurable 1-periodic function in the interval* [0, 1]. *Let* $S_3(f; x)$ *be its interpolating cubic spline, which interpolates f in the points* i/n, $i = 0, 1, \ldots, n$, *and satisfies the periodic conditions* (4.79). *Then*

$$\|f - S_3\|_p \leqslant c\tau_4(f; n^{-1})_p$$

where c is an absolute constant.

From the properties of τ_4 and Theorem 4.14 we obtain the following corollaries.

Corollary 4.10. *If the 1-periodic function f has first derivative* $f' \in L_p[0, 1]$, *then*

$$\|f - S_3\|_p = O[n^{-1}\omega_3(f'; n^{-1})_p]$$

Corollary 4.11. *If the 1-periodic function f has kth derivative* $f^{(k)}$, $k = 0, 1, 2, 3$, *of bounded variation in the interval* [0, 1], *then*

$$\|f - S_3\|_p = O(n^{-k-1}V_0^1 f^{(k)}) \quad k = 0, 1, 2, 3$$

In the case of the interpolating cubic spline satisfying the interpolation conditions (4.80) we can also obtain an estimate using $\tau_4(f; \delta)_p$, but we have an additional term depending on the boundary conditions. To achieve this we must apply a more complicated interpolation theorem than Theorem 2.9 (with boundary conditions). We shall not give the corresponding result here.

4.4. Notes

The Bernstein polynomials $B_n(f; x)$ approximate the function f at the ends of the interval [0, 1] better than in the middle of the interval. Using this fact Ivanov [5] has obtained an improvement of Theorem 4.1. He introduces weighted averaged moduli in the following way. Let $\psi(t, x)$ be a function of $t > 0$ and $x \in [0, 1]$. Let

$$\omega_k(f, x; \psi(t)) = \sup\left\{|\Delta_h^k f(y)| : y, y + kh \in \left[x - \frac{k\psi(t, x)}{2}, x + \frac{k\psi(t, x)}{2}\right]\right\}$$

$$\tau_k(f; \psi(t))_p = \|\omega_k(f, \cdot; \psi(t))\|_p$$

$$\Delta(t, x) = t\sqrt{x(1 - x)} + \frac{t^2}{2}$$

Then we have

$$\|f - B_n(f; \cdot)\|_p \leqslant c\tau_k(f; \Delta(n^{-1/2}))_p \tag{4.98}$$

where c is an absolute constant, $1 \leqslant p \leqslant \infty$.

Ivanov [8] has obtained the first inverse theorem for Bernstein operators in L_p, $1 < p < \infty$. We shall give here the main results.

Let $1 < p < \infty$, $1/p < \alpha < 1$, $f \in C[0, 1]$. Then the following assertions are equivalent:

(a) $\|f - B_n(f; \cdot)\|_p = O(n^{-\alpha})$ $n \to \infty$

(b) $\tau_2(f; \Delta(\delta))_p = O(\delta^{2\alpha})$ $\delta \to 0$

Let $1 < p < \infty$, $f \in C[0, 1]$. Then the following assertions are equivalent:

(a) $\|f - B_n(f; \cdot)\|_p = O(n^{-1})$ $n \to \infty$

(b) $\tau_2(f; \Delta(\delta))_p = O(\delta^2)$ $\delta \to 0$

(c) f is absolutely continuous in $[0, 1]$, f' is absolutely continuous in $(0, 1)$, the function $x(1 - x)f''(x)$ belongs to L_p.

Let $1 < p < \infty$, $f \in C[0, 1]$. Then $\|f - B_n(f; \cdot)\|_p = O(n^{-1})$ iff f is linear.

Van Wickeren [1] has obtained the following result which is a little stronger than (4.98) for $p = 1$:

$$\overline{\int} \sup_{k \geqslant n} |B_k(f; x) - f(x)| \, dx \leqslant M\tau_2(f; \Delta(n^{-1/2}))_p$$

where $\overline{\int}$ is the upper Riemann integral.

Theorem 4.3 is given by Popov [7]. For 2π-periodic functions the analogue of Theorem 4.3 is given by Hristov [5].

If

$$L(1, x) = 1, \; L(\sin t, x) = \sin x - \alpha(x), \; L(\cos t, x) = \cos x - \beta(x)$$

then

$$\|f - L(f; \cdot)\|_p \leqslant c\tau(f; \sqrt{d})_p \quad 1 \leqslant p \leqslant \infty,$$

where c is an absolute constant and

$$d = \|\alpha(x)\sin x + \beta(x)\cos x\|_{C[0,2\pi]} = \left\| L\left(2\sin^2\frac{x-t}{2}, x\right)\right\|_{C[0,2\pi]}$$

Mevissen *et al.* [1] give a generalization of Theorem 4.3 in terms of the upper Riemann integral.

The main results from Section 4.3 are obtained by Andreev and Popov [1]. The generalization of these results for splines of arbitrary degree (not only degree 1, 2 and 3) is obtained by Binev [3].

Andreev [3] obtained estimates for the derivatives of the difference between the function f and its quadratic or cubic interpolation spline. Using the notation of Section 4.3 his results are as follows:

$$\|f^{(i)} - S_2^{(i)}\|_p \leqslant 10n^{-2+i}\omega(f''; n^{-1})_p \qquad i = 0, 1, 2$$

$$\|f^{(i)} - S_2^{(i)}\|_p \leqslant 27n^{-1+i}\omega(f'; n^{-1})_p \qquad i = 0, 1$$

$$\|f^{(i)} - S_3^{(i)}\|_p \leqslant 29n^{-3+i}\omega(f'''; n^{-1})_p \qquad i = 0, 1, 2, 3$$

$$\|f^{(i)} - S_3^{(i)}\|_p \leqslant 71n^{-2+i}\omega(f''; n^{-1})_p \qquad i = 0, 1, 2$$

$$\|f^{(i)} - S_3^{(i)}\|_p \leqslant 154n^{-1+i}\omega(f'; n^{-1})_p \qquad i = 0, 1$$

These results were generalized by Binev [3] for interpolation splines of arbitrary degree. Briefly, his result is as follows. Let S_r be the interpolation spline of rth degree with equidistant knots of step $1/n$ for the 1-periodic function f. If the corresponding derivatives of f exist, then

$$\|f^{(i)} - S_r^{(i)}\|_p \leqslant c(r, v)n^{-v+i}\omega_{r+1-v}(f^{(v)}; n^{-1})_p \quad v = 1, \ldots, r, i = 0, \ldots, v$$

Binev [1], [2], [4] discovered the effect of superconvergence in spline interpolation. Briefly this means that there exist some points different from the interpolation points, for which the order of convergence is one degree better then the global order of convergence. These points are connected with the zeros of the Bernoulli polynomials and the order of convergence can be estimated by the averaged moduli of smoothness.

Hristov [3], [4] has considered the problem of convergence of interpolation trigonometric polynomials is an L_p-metric. We shall give here one of his results. Let

$$x_j^{(n)} = \frac{2\pi j}{2n+1} \quad j = 0, 1, \ldots, 2n$$

and let $I_n(f; x)$ be the interpolation trigonometric polynomial for the bounded 2π-periodic function f with respect to the points $x_j^{(n)}, j = 0, 1, \ldots, 2n$. Then for $1 < p < \infty$ we have $\|f - I_n(f; \cdot)\|_p \leqslant c(k, p)\tau_k(f; n^{-1})_p$.

Let

$$I_n(f; x) = \frac{a_0^{(n)}}{2} + \sum_{v=1}^{n} [a_v^{(n)}(f)\cos vx + b_v^{(n)}(f)\sin vx]$$

Hristov [2] shows that (for $1 \leqslant p \leqslant \infty$)

$$\max\{|a_v^{(n)}(f)|, |b_v^{(n)}(f)|\} \leqslant c(k)\tau_k\left(f; \frac{1}{v}\right)_p \quad v = 1, \ldots, n$$

Popov and Szabandos [1] have obtained estimates using the averaged second modulus of smoothness for the approximation of functions by means of the so-called discrete Jackson operator:

$$\mathcal{T}_n(f; x) = \frac{1}{n^2}\sum_{k=0}^{n-1} f(t_k)\frac{\sin^2\frac{n}{2}(x-t_k)}{\sin^2\frac{x-t_k}{2}} \quad t_k = \frac{2k\pi}{n}, k = 0, \ldots, n-1$$

In the same paper they solved the saturation problem for this operator in L_p, $1 \leqslant p < \infty$.

Drianov [6] considers the approximation of functions in $L_p(-\infty, \infty)$ by means of a sequence of discrete linear operators of exponential type. He considers the operators

$$I_\sigma(f; x) = \sum_{k=-\infty}^{\infty} f(t_k) \frac{\sin^2 \frac{\sigma}{2}(x - t_k)}{\sin \frac{x - t_k}{2}} \qquad t_k = \frac{2k\pi}{\sigma}, k = 0 \pm 1, \ldots$$

defined for a bounded function on $(-\infty, \infty)$. The following estimate is given. Let $f \in L_1 \cap L_\infty$, then

$$\| f - I_\sigma(f; \cdot) \|_1 \leqslant c \left[\frac{1}{\sigma} \int_{1/2}^\sigma \tau \left(f; \frac{1}{u} \right)_1 du + \tau \left(f; \frac{1}{\sigma} \right)_1 + \frac{\| f \|_{L_1}}{\sigma} \right]$$

In the same paper Drianov solves the saturation problem for these operators.

5 ESTIMATION OF THE ERROR IN THE NUMERICAL SOLUTION OF INTEGRAL EQUATIONS

In this chapter we shall apply the averaged moduli of smoothness to obtain estimates of the error in the numerical solution of the Fredholm integral equation of the second kind using either collocation splines of first, second and third degree or the method of replacing the integral with a quadrature formula. With small modifications the techniques applied can be used for estimating the error of the numerical solution of the Volterra integral equation of the second kind, but we shall not state the corresponding results here.

5.1. The method of collocation splines

Let us consider the Fredholm integral equation of the second kind

$$y(x) - \lambda \int_0^1 K(x,t)y(t)\,\mathrm{d}t = f(x) \quad x \in [0,1] \tag{5.1}$$

We shall assume that equation (5.1) has a unique solution, i.e. that λ is not an eigenvalue. We shall consider numerical solutions of equation (5.1) by means of collocation linear, quadratic and cubic splines. In the second and third cases we shall restrict ourselves only to the periodic case, so in these cases we shall assume that the kernel K and the term in the right-hand side of f are 1-periodic functions of their arguments. As a consequence, the solution in these cases will also be a 1-periodic function.

As in Section 4.3, let us set $x_i = ih$, $h = 1/n$, $i = -2, -1, 0, \ldots, n+2$, $B_{j,k}(x) =$

$B_{j,0}(x - kh)$

$$S_1(x) = \sum_{k=0}^{n} a_k B_{1,k}(x), \quad B_{1,0}(x) = n[(x+h)_+ - 2x_+ + (x-h)_+] \tag{5.2}$$

$$S_2(x) = \sum_{k=0}^{n+1} c_k B_{2,k}(x), \quad B_{2,0}(x) = \frac{n^2}{2} \sum_{k=0}^{3} (-1)^k \binom{3}{k} \left(x + \frac{3h}{2} - kh\right)_+^2 \tag{5.3}$$

$$S_3(x) = \sum_{k=-2}^{n+2} d_k B_{3,k}(x); \quad B_{3,0}(x) = \frac{n^3}{6} \sum_{k=0}^{4} (-1)^k \binom{4}{k} (x + 2h - kh)_+^3 \tag{5.4}$$

We shall find the numerical solution of (5.1) in the form of equations (5.2), (5.3) and (5.4) respectively. The coefficients a_k, $k = 0,\ldots,n$, c_k, d_k, $k = -1, 0,\ldots,n+1$, will be determined from the collocation conditions at the points x_i, $i = 0, 1,\ldots,n$;

$$S_k(x_i) = f(x_i) + \lambda \int_0^1 K(x_i, t) S_k(t)\, dt \quad i = 0,\ldots,n, \; k = 1, 2, 3 \tag{5.5}$$

For the quadratic and cubic splines we shall also require the necessary boundary conditions for the periodicity

$$S_k^{(i)}(0) = S_k^{(i)}(1) \quad i = 1, 2, k = 2, 3 \tag{5.6}$$

to be satisfied.

Since S_k is linear with respect to the coefficients

$$a_k, \; k = 0, 1,\ldots,n, \; c_k, \; d_k, \; k = -1, 0,\ldots,n+1,$$

the equations (5.5) together with the boundary conditions (5.6) represent three systems of linear algebraic equations with respectively $n+1$, $n+3$ and $n+3$ unknowns. We shall assume that these systems have unique solutions.

Our aim will be to find an estimate for

$$\| f - S_k \|_{L_p[0,1]} \quad k = 1, 2, 3, \; 1 \leqslant p \leqslant \infty$$

Let $S_k(y; x)$, $k = 1, 2, 3$, be the interpolation spline of degree k, $k = 1, 2, 3$, which interpolates the solution y at the points x_i and which satisfies for $k = 2, 3$ the corresponding periodic boundary conditions (see Section 4.3);

$$S_k^{(i)}(y; 0) = S_k^{(i)}(y; 1) \quad i = 1, 2, \; k = 2, 3 \tag{5.7}$$

From equation (5.1), in view of

$$y(x_i) = S_k(y; x_i) \quad i = 0,\ldots,n$$

we obtain

$$S_k(y; x_i) = f(x_i) + \lambda \int_0^1 K(x_i, t) S_k(y; t)\, dt + \lambda \int_0^1 K(x_i, t)[y(t) - S_k(y; t)]\, dt \tag{5.8}$$

Let us set $\varphi_k = S_k(y; \cdot) - S_k$, where S_k is the collocation spline which satisfies the collocation conditions (5.5) and (for $k = 2, 3$) the boundary conditions (5.6).

If we subtract equations (5.5) from (5.8) we obtain

$$\varphi_k(x_i) = \lambda \int_0^1 K(x_i,t)\varphi_k(t)\,dt + \lambda \int_0^1 K(x_i,t)[y(t) - S_k(y,t)]\,dt \quad \text{for } i = 0,1,\ldots,n.$$
(5.9)

Notice that for every spline S_k of kth degree, $k = 1,2,3$, which has the representation (5.2)–(5.4), (5.6), we have

$$\|S_k\|_{C[0,1]} \leqslant M_k \max_{0 \leqslant i \leqslant n} |S_k(x_i)|$$
(5.10)

where the constant M_k depends only on k. Obviously $M_1 = 1$. From Lemma 4.7, Corollary 4.4 we obtain $M_2 \leqslant 5$, and from Lemma 4.8, Corollary 4.10 we obtain $M_3 \leqslant 75/4$. It is not difficult to see from the proof of Lemma 4.8 that if we work more precisely the constant $75/4$ can be replaced by $75/12$ and therefore $M_3 \leqslant 7$. If we apply (5.10) to the spline φ_k (which also has the form (5.2)–(5.4) and satisfies the periodic boundary conditions, because the same is valid for $S_k(y; x)$ and $S_k(x)$) we obtain

$$\frac{1}{M_k}\|\varphi_k\|_{C[0,1]} \leqslant \max\{|\varphi_k(x_i)|: 0 \leqslant i \leqslant n\} \quad k = 1,2,3$$
(5.11)

Suppose now that λ and the kernel K satisfy the inequality

$$|\lambda| \sup_{0 \leqslant x \leqslant 1} \int_0^1 |K(x,t)|\,dt = \rho_k < \frac{1}{M_k}$$
(5.12)

Then equations (5.9) give us

$$|\varphi_k(x_i)| \leqslant \rho_k \|\varphi_k\|_{C[0,1]} + |\lambda| \|K\|_{C,q} \|y - S_k(y;\cdot)\|_{L_p}$$
(5.13)

where $i = 0,1,\ldots,n$, $k = 1,2,3$, and where we have set

$$\|K\|_{C,q} = \sup\{\|K(x;\cdot)\|_{L_q}: 0 \leqslant x \leqslant 1\} \quad \frac{1}{p} + \frac{1}{q} = 1$$

The inequalities (5.13) and (5.11) together with condition (5.12) give us

$$\|\varphi_k\|_{C[0,1]} = \frac{M_k|\lambda|\|K\|_{C,q}\|y - S_k(y;\cdot)\|_{L_p}}{1 - \rho_k M_k}.$$
(5.14)

Therefore

$$\|y - S_k\|_{L_p} \leqslant \|y - S_k(y;\cdot)\|_{L_p} + \|\varphi_k\|_{L_p}$$
$$\leqslant \left(1 + \frac{M_k|\lambda|\|K\|_{C,q}}{1 - \rho_k M_k}\right)\|y - S_k(y;\cdot)\|_{L_p}$$

We have obtained the following theorem.

Theorem 5.1. *For the approximate solution S_k of equation (5.1), obtained by the collocation method (5.5), (5.6), we have, if condition (5.12) is satisfied,*

$$\|y - S_k\|_{L_p[0,1]} \leqslant c(\lambda, q, K, k)\|y - S_k(y)\|_{L_p[0,1]} \quad k = 1,2,3$$

where $c(\lambda, q, K, k) \leqslant (1 + M_k|\lambda| \|K\|_{c,q}/(1 - \rho_k M_k))$, and $S_k(y)$ is the interpolation spline of kth degree for the solution y, with interpolation knots $x_i = i/n$, $i = 0, 1, \ldots, n$, and the periodic conditions (5.7).

In view of the estimates obtained in Section 4.3 for the interpolation splines (periodic) of first, second and third degree using the averaged moduli of smoothness, we obtain the following.

Theorem 5.2. *If condition (5.12) holds, then*

$$\|y - S_k\|_{L_p} = O(\tau_{k+1}(y; n^{-1})_{L_p}) \quad k = 1, 2, 3$$

Using the properties of the averaged moduli of smoothness, we can obtain many consequences from Theorem 5.2. We shall give only the following.

Corollary 5.1. *If the solution y has a derivative y', $y' \in L_p[0, 1]$, then*

$$\|y - S_k\|_{L_p} = O[n^{-1}\omega_k(y'; n^{-1})_{L_p}] \quad k = 1, 2, 3$$

If the solution y has derivatives $y^{(i)}$ of bounded variation, $i = 0, \ldots, k$, then

$$\|y - S_k\|_{L_p} = O(n^{-i-1})$$

Remark. The properties of y are determined by the properties of the kernel K and the right-hand side of f. Hence the estimates of the above theorem can be expressed by the averaged moduli of smoothness of K and f. We shall not give such estimates here.

The estimate (5.14) gives us the error in the collocation points x_i, $i = 0, 1, \ldots, n$.

Theorem 5.3. *Under condition (5.12) for the error at the collocation points we have*

$$|y(x_i) - S_k(x_i)| \leqslant \frac{M_k|\lambda| \|k\|_{c,q}}{1 - \rho_k M_k} \|y - S_k(y; \cdot)\|_{L_p}, \quad k = 1, 2, 3$$

Corollary 5.2. *Under condition (5.12) for the error at the collocation points we have*

$$|y(x_i) - S_k(x_i)| = O[\tau_{k+1}(y; n^{-1})_{L_p}] \quad k = 1, 2, 3$$

5.2. A general estimate for the approximate solution in integral norms

In this section we shall give an estimate for the L_p-norm of the difference of the solutions of two Fredholm integral equations of the second kind using the difference of the corresponding kernels and right-hand sides. Such an estimate in the uniform norm ($p = \infty$) is well known.

Theorem 5.4.   *Given the integral equations*

$$y(x) - \lambda \int_a^b K(x,t)y(t)\,dt = f(x) \quad x \in [a,b] \tag{5.15}$$

$$\tilde{y}(x) - \lambda \int_a^b \tilde{K}(x,t)\tilde{y}(t)\,dt = \tilde{f}(x) \quad x \in [a,b] \tag{5.16}$$

let $R(x,t;\lambda)$ be the resolvant of equation (5.15), i.e. let the solution of (5.15) have the form

$$y(x) = f(x) + \lambda \int_a^b R(x,t;\lambda)f(t)\,dt$$

Let the following conditions be satisfied:

(i)
$$\left\{ \int_a^b \left[\int_a^b |R(x,t;\lambda)|^q\,dt \right]^{p/q} dx \right\}^{1/p} = R_{p,q} < \infty$$

(ii)
$$\left\{ \int_a^b \left[\int_a^b |K(x,t) - \tilde{K}(x,t)|^q\,dt \right]^{p/q} dx \right\}^{1/p} = K_{p,q} < \infty$$

$$\frac{1}{p} + \frac{1}{q} = 1$$

$$1 - |\lambda| K_{p,q}(1 + |\lambda| R_{p,q}) > 0 \tag{5.17}$$

Then the following estimate holds:

$$\|y - \tilde{y}\|_{L_p} \leqslant (1 + |\lambda| R_{p,q}) \left[\|f - \tilde{f}\|_{L_p} + \frac{|\lambda| K_{p,q}(1 + |\lambda| R_{p,q}) \|\tilde{f}\|_{L_p}}{1 - |\lambda| K_{p,q}(1 + |\lambda| R_{p,q})} \right]$$

Proof.   From (5.15) and (5.16) we get

$$y(x) - \tilde{y}(x) = \lambda \left[\int_a^b K(x,t)y(t)\,dt - \int_a^b \tilde{K}(x,t)\tilde{y}(t)\,dt \right] + f(x) - \tilde{f}(x)$$

$$= \lambda \int_a^b K(x,t)[y(t) - \tilde{y}(t)]\,dt + f(x) - \tilde{f}(x) + \lambda \int_a^b [K(x,t) - \tilde{K}(x,t)]\tilde{y}(t)\,dt$$

i.e. for $\varepsilon = y - \tilde{y}$ we have the equation

$$\varepsilon(x) - \lambda \int_a^b K(x,t)\varepsilon(t)\,dt = g(x) \tag{5.18}$$

where

$$g(x) = \lambda \int_a^b [K(x,t) - \tilde{K}(x,t)]\tilde{y}(t)\,dt + f(x) - \tilde{f}(x)$$

Equation (5.18) has the same resolvant $R(x,t;\lambda)$ as equation (5.15) and its solution is represented in the form

$$\varepsilon(x) = g(x) + \lambda \int_a^b R(x,t;\lambda)g(t)\,dt$$

From here we obtain

$$\|y - \tilde{y}\|_{L_p[a,b]} = \|\varepsilon\|_{L_p}$$
$$\leqslant \|g\|_{L_p} + |\lambda| R_{p,q} \|g\|_{L_p}$$
$$= (1 + |\lambda| R_{p,q}) \|g\|_{L_p} \tag{5.19}$$

where

$$R_{p,q} = \left\{ \int_a^b \left[\int_a^b |R(x,t;\lambda)|^q \, dt \right]^{p/q} dx \right\}^{1/p}$$

We have used Hölder's inequality

$$\left\| \int_a^b R(x,t;\lambda) g(t) \, dt \right\|_{L_p[a,b]} \leqslant R_{p,q} \|g\|_{L_p[a,b]}$$

Therefore, to obtain an estimate for $\|\varepsilon\|_{L_p}$, we must estimate $\|g\|_{L_p}$. Applying Hölder's inequality again, we obtain

$$\|g\|_{L_p[a,b]} \leqslant |\lambda| \left\{ \int_a^b \left| \int_a^b [K(x,t) - \tilde{K}(x,t)] \tilde{y}(t) \, dt \right|^p dx \right\}^{1/p} + \|f - \tilde{f}\|_p$$
$$\leqslant |\lambda| \left\{ \int_a^b \left[\int_a^b |K(x,t) - \tilde{K}(x,t)|^q \, dt \right]^{p/q} dx \right\}^{1/p} \|\tilde{y}\|_{L_p} + \|f - \tilde{f}\|_p$$
$$= |\lambda| K_{p,q} \|\tilde{y}\|_p + \|f - \tilde{f}\|_p \tag{5.20}$$

It remains to estimate $\|\tilde{y}\|_p$. From (5.16), by subtracting

$$\lambda \int_a^b K(x,t) \tilde{y}(t) \, dt$$

from both sides of the equation, we obtain

$$\tilde{y}(x) - \lambda \int_a^b K(x,t) \tilde{y}(t) \, dt = \lambda \int_a^b [\tilde{K}(x,t) - K(x,t)] \tilde{y}(t) \, dt + \tilde{f}(x)$$

Again, using the resolvent of equation (5.15), we find that

$$\tilde{y}(x) = \varphi(x) + \lambda \int_a^b R(x,t;\lambda) \varphi(t) \, dt \tag{5.21}$$

where

$$\varphi(x) = \tilde{f}(x) + \lambda \int_a^b [\tilde{K}(x,t) - K(x,t)] \tilde{y}(t) \, dt$$

From (5.21) we obtain, as in (5.18) and (5.19),

$$\|\tilde{y}\|_p \leqslant \|\varphi\|_p + |\lambda| R_{p,q} \|\varphi\|_p \tag{5.22}$$

The norm of φ in L_p can be estimated, as in the case of $\|g\|_{L_p}$, more exactly

$$\|\varphi\|_p \leqslant \|\tilde{f}\|_p + |\lambda| K_{p,q} \|\tilde{y}\|_p \tag{5.23}$$

From (5.22) and (5.23) we obtain

$$\|\tilde{y}\|_{L_p} \leqslant \|\tilde{f}\|_{L_p} + |\lambda| K_{p,q} \|\tilde{y}\|_{L_p} + |\lambda| R_{p,q} \|\tilde{f}\|_{L_p} + \lambda^2 R_{p,q} K_{p,q} \|\tilde{y}\|_{L_p}$$

In view of condition (5.17), the above inequality gives

$$\|\tilde{y}\|_{L_p} \leqslant \frac{(1 + |\lambda| R_{p,q}) \|\tilde{f}\|_{L_p}}{1 - |\lambda| K_{p,q}(1 + |\lambda| R_{p,q})} \tag{5.24}$$

If we substitute the estimate (5.24) into (5.20) and (5.20) into (5.19), we obtain the required estimate

$$\|\varepsilon\|_{L_p} \leqslant (1 + |\lambda| R_{p,q}) \left\{ \|f - \tilde{f}\|_{L_p} + \frac{|\lambda| K_{p,q}(1 + |\lambda| R_{p,q}) \|f\|_{L_p}}{1 - |\lambda| K_{p,q}(1 + |\lambda| R_{p,q})} \right\} \qquad \blacksquare$$

Although the estimate of Theorem 5.4 is very rough, we can apply it in the cases when we wish to estimate the order of $\|y - \tilde{y}\|_{L_p}$.

5.3. The method of replacing the integral by a quadrature formula

The method of replacing the integral by a quadrature formula for the Fredholm integral equation of the second kind

$$y(x) = \lambda \int_0^1 K(x,t)y(t)\, dt + f(x) \quad x \in [0,1] \tag{5.25}$$

is as follows. We take some quadrature formula

$$\int_0^1 g(t)\, dt = \sum_{i=1}^n A_i g(x_i) + R_n(g) \quad x_i \in [0,1], \ i = 1, 2, \dots, n \tag{5.26}$$

we substitute the integral in equation (5.25) by the quadrature formula (5.26), ignoring the error R_n. Substituting $x = x_j, \ j = 1, 2, \dots, n$, we obtain

$$\tilde{y}(x_j) = \lambda \sum_{i=1}^n A_i K(x_j, x_i)\tilde{y}(x_i) + f(x_j) \quad j = 1, \dots, n \tag{5.27}$$

Equations (5.27) represent a system of n linear algebraic equations for the unknowns $\tilde{y}(x_j), \ j = 1, 2, \dots, n$ ($\tilde{y}(x)$ is the approximate solution).

Let us consider in detail one particular case – the rectangle rule:

$$\int_0^1 g(t)\, dt = \frac{1}{n} \sum_{i=1}^n g\left(\frac{2i-1}{2n}\right) + R_n(g) \tag{5.28}$$

Then (5.27) takes the form

$$\tilde{y}\left(\frac{2j-1}{2n}\right) = \frac{\lambda}{n} \sum_{i=1}^n K\left(\frac{2j-1}{2n}, \frac{2i-1}{2n}\right) \tilde{y}\left(\frac{2i-1}{2n}\right) + f\left(\frac{2j-1}{2n}\right) \quad j = 1, \dots, n \tag{5.29}$$

using the notation

$$\tilde{y}\left(\frac{2i-1}{2n}\right) = \tilde{y}_i$$

$$K\left(\frac{2j-1}{2n}, \frac{2i-1}{2n}\right) = K_{j,i}$$

$$f\left(\frac{2i-1}{2n}\right) = f_i$$

System (5.29) then becomes

$$\tilde{y}_j - \frac{\lambda}{n} \sum_{i=1}^{n} K_{ji} \tilde{y}_i = f_j \quad j = 1, 2, \ldots, n \tag{5.30}$$

Equations (5.30) can be considered as replacing the right-hand side of f and the kernel K of equation (5.25) by another right-hand side, $\tilde{f}$, and another (degenerate) kernel, $\tilde{K}$. In fact, let us set

$$\tilde{K}(x, t) = K(x_k, x_i) \quad \text{if } x \in \left[\frac{k-1}{n}, \frac{k}{n}\right), \ t \in \left[\frac{i-1}{n}, \frac{i}{n}\right)$$

$$\tilde{f}(x) = f(x_k) \qquad \text{if } x \in \left[\frac{k-1}{n}, \frac{k}{n}\right) \tag{5.31}$$

$$x_k = \frac{2k-1}{2n} \quad k = 1, 2, \ldots, n$$

(for $k = n$ or for $i = n$ we take the closed interval $[(n-1)/n, 1]$.)

Let us consider the integral equation

$$\tilde{y}(x) - \lambda \int_0^1 \tilde{K}(x, t) \tilde{y}(t) \, dt = \tilde{f}(x) \quad x \in [0, 1] \tag{5.32}$$

Since the kernel $\tilde{K}$ and the right-hand side, $\tilde{f}$, are constants in the interval $\Delta_k = [(k-1)/n, k/n)$, the solution $\tilde{y}$ is also a constant in this interval; let its values be $\tilde{y}_k$. Indeed, for $x \in [(k-1)/n, k/n)$ we have

$$\tilde{y}(x) = \lambda \sum_{i=1}^{n} K(x_k, x_i) \int_{(i-1)/n}^{i/n} \tilde{y}(t) \, dt + f(x_k)$$

and $\tilde{y}$ is constant in the interval Δ_k, i.e. $\tilde{y}(x) = \tilde{y}_k$, $x \in \Delta_k$, and equation (5.32) evidently takes the form of (5.30). Therefore for the estimate of the difference $y - \tilde{y}$ between the solutions of equation (5.25) and equation (5.30) we can apply the general theorem of Section 5.2. We shall assume that the resolvant $R(x, t; \lambda)$ of equation (5.25) satisfies the condition

$$\left\{ \int_0^1 \left[\int_0^1 |R(x, t; \lambda)|^q \, dt \right]^{p/q} dx \right\}^{1/p} = R_{p,q} < \infty \tag{5.33}$$

Theorem 5.5. *Let K and f be bounded measurable functions. For the difference $\varepsilon = y - \tilde{y}$ between the solutions of equations (5.25) and (5.30) using assumption (5.33) and*

$$1 - |\lambda|(1 + |\lambda| R_{p,q})\tau_t(K; n^{-1})_{p,q} > 0 \tag{5.34}$$

we have

$$\|\varepsilon\|_{L_p} \leqslant (1 + |\lambda| R_{p,q}) \left[\tau(f; n^{-1})_{L_p} + \frac{|\lambda|(1 + |\lambda| R_{p,q})\|f\|_C \tau_t(K; n^{-1})_{p,q}}{1 - |\lambda|(1 + |\lambda| R_{p,q})\tau_t(K; n^{-1})_{p,q}} \right]$$

where

$$\tau_t(K; \delta)_{p,q} = \|\tau_t(K(\cdot, t); \delta)_{L_q}\|_{L_p}$$

$$\tau_t(K(x, t); \delta)_{L_q} = \|\omega_t(K(x, t), \cdot; \delta)\|_{L_q}$$

$$\omega_t(K(x, t), z; \delta) = \sup\left\{ |K(x, t') - K(x, t'')| : |t' - z| \leqslant \frac{\delta}{2}, |t'' - z| \leqslant \frac{\delta}{2} \right\} \quad \frac{1}{p} + \frac{1}{q} = 1$$

Proof. To apply Theorem 5.4 we must estimate $K_{p,q}$ and $\|f - \tilde{f}\|_{L_p}$. Using (5.31) we obtain

$$K_{p,q} = \left\{ \int_0^1 \left[\int_0^1 |K(x, t) - \tilde{K}(x, t)|^q \, dt \right]^{p/q} dx \right\}^{1/p}$$

$$= \left\{ \int_0^1 \left[\sum_{i=1}^n \int_{(i-1)/n}^{i/n} |K(x, t) - K(x, x_i)|^q \, dt \right]^{p/q} dx \right\}^{1/p}$$

$$\leqslant \left\{ \int_0^1 \left[\sum_{i=1}^n \int_{(i-1)/n}^{i/n} (\omega_{t'}(K(x, t'), t; n^{-1})^q \, dt \right]^{p/q} dx \right\}^{1/p}$$

$$= \left\{ \int_0^1 [\|\omega_{t'}(K(x, t'), \cdot; n^{-1})\|_{L_q}]^p \, dx \right\}^{1/p} = \tau_t(K; n^{-1})_{p,q}$$

Using (5.31) we get

$$\|f - \tilde{f}\|_{L_p} = \left[\int_0^1 |f(x) - \tilde{f}(x)|^p \, dx \right]^{1/p} = \left[\sum_{i=1}^n \int_{(i-1)/n}^{i/n} |f(x) - f(x_i)|^p \, dx \right]^{1/p}$$

$$\leqslant \left[\sum_{i=1}^n \int_{(i-1)/n}^{i/n} (\omega(f, x; n^{-1}))^p \, dx \right]^{1/p} = \tau(f; n^{-1})_{L_p}$$

From the above two inequalities, in view of

$$\|\tilde{f}\|_{L_p} = \left[\sum_{i=1}^n |f(x_i)|^p n^{-1} \right]^{1/p} \leqslant \|f\|_C$$

and Theorem 5.4, we obtain the statement of Theorem 5.5. ∎

We can obtain several corollaries from Theorem 5.5, if we make some assumptions about the functions K and f. We shall give only two of them.

Corollary 5.3. *If the right-hand side, f, and the kernel, K, satisfy the Lipschitz* condition (*uniformly with respect to x for the kernel K*), *and if* $R_{\infty,1} < \infty$, *then*

$$\|y - \tilde{y}\|_{C[0,1]} = O(n^{-1})$$

We obtain this corollary immediately from Theorem 5.5 because, under the assumptions we have made, we have

$$\tau_t(K; n^{-1})_{\infty,1} = O(n^{-1})$$
$$\tau(f; n^{-1})_{L_\infty} = \omega(f; n^{-1}) = O(n^{-1})$$

On the other hand, for sufficiently large n, we have

$$1 - |\lambda|(1 + |\lambda| R_{\infty,1})\tau_t(K; n^{-1})_{\infty,1} > 0$$

and therefore we can apply Theorem 5.5.

Corollary 5.4. *Let f be a function of bounded variation in the interval* $[0, 1]$ *and let the kernel* $K(x,t)$, *for every fixed t, be a function of bounded variation with variation* $\leq V$ (*uniformly with respect to x*). *Let*

$$\int_0^1 \max_{0 \leq t \leq 1} |R(x, t; \lambda)| \, dx < \infty$$

Then

$$\|y - \tilde{y}\|_{L[0,1]} = O(n^{-1})$$

Again this corollary follows directly from Theorem 5.5, since from the assumptions we have made we have

$$\tau_t(K; n^{-1})_{1,\infty} = O(n^{-1})$$
$$\tau(f; n^{-1})_L = O(n^{-1})$$

and for sufficiently large n condition (5.34) holds.

This technique can be applied with success also in the other cases, when we solve the Fredholm integral equation of the second kind numerically, since in all practical cases these methods are equivalent to a method consisting of replacing the kernel and the right-hand side with another kernel and another right-hand side.

In the concrete case of the kernel and the right-hand side given by (5.31) and (5.32), we can also estimate $y - \tilde{y}$, using the method of Section 5.1.

The exact values of the solution $y(x_i)$ satisfy the equations

$$y(x_i) = \tilde{y}(x_i) = f(x_i) + \lambda \int_0^1 K(x_i, t)\tilde{y}(t) \, dt + \lambda \int_0^1 K(x_i, t)[y(t) - \tilde{y}(t)] \, dt \quad (5.35)$$

where $\tilde{y}(t) = \tilde{y}(x_i)$ for $t \in \Delta_i$.

From (5.35) and (5.30) we obtain

$$\tilde{y}(x_i) - y(x_i) = \lambda \int_0^1 K(x_i, t)[y(t) - \tilde{y}(t)] \, dt + \lambda \int_0^1 K(x_i, t)[\tilde{y}(t) - \tilde{y}(t)] \, dt$$

$$+ \lambda \sum_{j=1}^n \tilde{y}(x_j) \int_{(j-1)/n}^{j/n} [K(x_i, t) - K(x_i, x_j)] \, dt$$

From this we derive

$$\max \{|\tilde{y}(x_i) - \bar{y}(x_i)|: 1 \leqslant i \leqslant n\} \leqslant |\lambda| \, \|K\|_C \tau(y; n^{-1})_L + |\lambda| \, \|\tilde{y} - \bar{y}\|_C \|K\|_C$$
$$+ |\lambda| \, \|\tilde{y}\|_C \tau_t(K; n^{-1})_{\infty,1} \qquad (5.36)$$

since

$$\left| \lambda \int_0^1 K(x,t)[y(t) - \bar{y}(t)] \, dt \right| \leqslant |\lambda| \, \|K\|_C \sum_{i=1}^n \int_{(i-1)/n}^{i/n} |y(t) - y(x_i)| \, dt$$

$$\leqslant |\lambda| \, \|K\|_C \sum_{i=1}^n \int_{(i-1)/n}^{i/n} \omega(y, t; n^{-1}) \, dt$$

$$= |\lambda| \, \|K\|_C \tau(y; n^{-1})_L \sum_{j=1}^n \int_{(j-1)/n}^{j/n} |K(x,t) - K(x_i, x_j)| \, dt$$

$$\leqslant \sum_{j=1}^n \int_{(j-1)/n}^{j/n} \omega_t(K(x_i, t), \cdot; n^{-1}) \, dt$$

$$\leqslant \tau_t(K; n^{-1})_{\infty,1}.$$

In view of

$$\|\tilde{y}\|_C = \|\tilde{y} - \bar{y}\|_C + \|\bar{y}\|_C \leqslant \max_{1 \leqslant i \leqslant n} |\tilde{y}(x_i) - y(x_i)| + \|y\|_C$$

we obtain the following theorem from (5.36).

Theorem 5.6. *Using the previous notation, if*

$$1 - |\lambda| \, \|K\|_C - |\lambda| \tau_t(K; n^{-1})_{\infty,1} > 0$$

then we have the estimate

$$\max \{|y(x_i) - \tilde{y}(x_i)|: 1 \leqslant i \leqslant n\} \leqslant \frac{|\lambda|\{\|K\|_C \tau(y; n^{-1})_C + \|y\|_C \tau_t(K; n^{-1})_{\infty,1}}{1 - |\lambda| \, \|K\|_C - \lambda \tau_t(K; n^{-1})_{\infty,1}}$$

As a consequence of Theorem 5.6 we obtain a reinforcement of Theorem 5.5, Corollary 5.3, as follows.

Corollary 5.5. *Let the function f be of bounded variation on the interval* [0, 1] *and let the kernel K(x, t) be a function of bounded variation* $\leqslant V$, *with respect to x for every t, and with respect to t for every x. If we have* $1 - |\lambda| \, \|K\|_C > 0$, *then*

$$\max \{|y(x_i) - \tilde{y}(x_i)|: 1 \leqslant i \leqslant n\} = O(n^{-1})$$

It is easy in fact to verify that under the conditions of the corollary the function y (the solution of equation (5.25)) is also a function of bounded variation. Thus $\tau(y; \delta)_L = O(\delta)$ and we have also $\tau_t(K; n^{-1})_{\infty,1} = O(n^{-1})$. Therefore for sufficiently large n we have

$$1 - |\lambda| \, \|K\|_C - |\lambda| \tau_t(K; n^{-1})_{\infty,1} > 0$$

and we can apply Theorem 5.6.

The interesting fact, that at the points x_i we obtain the estimate $O(n^{-1})$ under weaker assumptions, is analogous to Theorem 5.3 and its corollary – the error in the points of interpolation has order $O(n^{-1})$ under weaker assumptions.

5.4. Notes

The results of Section 5.1 are obtained by Andreev [2]. In the same paper Andreev also gives estimates for the derivatives of the difference between the solution y of the Fredholm integral equation of the second kind and the approximate spline solutions. Using the notation of Section 5.1 we have the following results.

Under condition (5.12) we have

$$\| y^{(r)} - S_k^{(r)} \|_p \leqslant c(k,r)\omega_{k+1-r}(y^{(r)}; n^{-1})_p \quad r = 1,\ldots,k, \; k = 1,2,3$$

The results from Section 5.3 are not published.

Zhidkov *et al.* [1] consider the problem of the Gibbs effect for the spline collocation method for the numerical solution of the Fredholm integral equation of the second kind.

If the solution y is a piecewise continuous function, then at the points of discontinuity of the first kind there is a Gibbs effect of about 10 per cent for the quadratic spline collocation, and about the same for the cubic spline collocation.

6 ESTIMATION OF THE ERROR IN THE NUMERICAL SOLUTION OF THE CAUCHY PROBLEM FOR ORDINARY DIFFERENTIAL EQUATIONS

In this chapter we shall show how it is possible to obtain estimates, using the averaged moduli of smoothness, of the error in the numerical solution of the Cauchy problem for ordinary differential equations using the Runge–Kutta and Adams-Bashforth methods. In principle it is possible to estimate the error of every difference (discrete) method by means of the averaged moduli of smoothness. We shall restrict ourselves only to the simplest cases, to illustrate the ideas more clearly.

6.1. Runge–Kutta methods

Let us consider the initial value problem:

$$y' = f(x, y) \quad x \in [0, A], \; A > 0$$
$$y(0) = y_0 \tag{6.1}$$

We shall assume that the right-hand side satisfies a Lipschitz condition with respect to the variable y, i.e.

$$|f(x, y) - f(x, z)| \leqslant K|y - z| \tag{6.2}$$

where K is an absolute constant.

Let us first consider Euler's method, which is the simplest method of the Runge–Kutta type. As is well known, in this method the approximate values

117

of the solution at the points $x_i = ih$, $h = A/n$, $i = 0, 1, \ldots, n$, are obtained by the formulae

$$\tilde{y}_{i+1} = \tilde{y}_i + hf(x_i, \tilde{y}_i) \quad i = 0, 1, \ldots, n - 1 \tag{6.3}$$

We estimate the error in the ith step by means of the error in the $(i - 1)$th step. From (6.1)–(6.3) we obtain

$$|y_{i+1} - \tilde{y}_{i+1}| = |y_{i+1} - \tilde{y}_i - hf(x_i, \tilde{y}_i) + hf(x_i, y_i) - hf(x_i, y_i)|$$

$$\leqslant |y_{i+1} - \tilde{y}_i + y_i - y_i - hf(x_i, y_i)| + Kh|y_i - \tilde{y}_i|$$

$$\leqslant (1 + hK)(y_i - \tilde{y}_i) + h\left|\frac{y_{i+1} - y_i}{h} - y_i'\right|$$

Since

$$\left|\frac{y_{i+1} - y_i}{h} - y_i'\right| \leqslant \omega(y', x_{i+1/2}; h); \left(x_{i+1/2} = x_i + \frac{h}{2}\right)$$

then

$$|y_{i+1} - \tilde{y}_{i+1}| \leqslant (1 + hK)|y_i - \tilde{y}_i| + h\omega(y', x_{i+1/2}; h) \tag{6.4}$$

If we apply the inequality (6.4) recursively, we obtain

$$|y_{i+1} - \tilde{y}_{i+1}| \leqslant (1 + hK)|y_i - \tilde{y}_i| + h\omega(y', x_{i+1/2}; h)$$

$$\leqslant (1 + hK)^2|y_{i-1} - \tilde{y}_{i-1}| + (1 + hK)h\omega(y', x_{i-1/2}; h) + h\omega(y', x_{i+1/2}; h)$$

$$\leqslant \cdots \leqslant (1 + hK)^i h\omega(y', x_{1/2}; h) + (1 + hK)^{i-1}h\omega(y', x_{3/2}; h) + \cdots$$

$$+ h\omega(y', x_{i+1/2}; h)$$

From the last inequality it follows that

$$|y_{j+1} - \tilde{y}_{j+1}| \leqslant (1 + hK)^n \sum_{i=0}^{n-1} h\omega(y', x_{i+1/2}; h)$$

$$= \left(1 + \frac{AK}{n}\right)^n \sum_{i=0}^{n-1} \int_{x_i}^{x_{i+1}} \omega(y', x_{i+1/2}; h) \, dx$$

$$\leqslant e^{KA} \sum_{i=0}^{n-1} \int_{x_i}^{x_{i+1}} \omega(y', x; 2h) \, dx$$

$$= e^{KA} \int_0^A \omega(y', x; 2h) \, dx$$

$$= A e^{KA} \tau(y'; 2h)_{L[0,A]}$$

So we have obtained the following statement.

Theorem 6.1. *For the numerical solution of problem* (6.1) *using assumption* (6.2) *by means of Euler's method* (6.3) *the following estimate holds:*

$$\max\{|y_i - \tilde{y}_i| : 0 \leqslant i \leqslant n\} \leqslant 2A e^{KA} \tau(y'; h)_{L[0,A]}$$

It is well known that Euler's method is a Runge–Kutta method with a local error $O(h^2)$. Now we shall obtain an estimate for those Runge–Kutta methods which have a local error $O(h^3)$. These methods are given by the formulae

$$\tilde{y}_{i+1} = \tilde{y}_i + phf(x_i, \tilde{y}_i) + qhf(x_i + \alpha h, \tilde{y}_i + \beta hf(x_i, \tilde{y}_i))$$

$$\tilde{y}_0 = y_0, \quad h = \frac{A}{n}, \quad i = 0, 1, \ldots, n-1 \tag{6.5}$$

where the constants p, q, α, β satisfy the system

$$p + q = 1$$
$$q\alpha = \tfrac{1}{2} \tag{6.6}$$
$$q\beta = \tfrac{1}{2}$$

From (6.6) it follows that this system has a one-parameter solution of the form

$$p = 1 - s, \quad q = s, \quad \alpha = \beta = \frac{1}{2s} \tag{6.7}$$

From (6.1), (6.5) and (6.7) we obtain

$$|y_{i+1} - \tilde{y}_{i+1}| = |y_{i+1} - \tilde{y}_i - phf(x_i, \tilde{y}_i) - qhf(x_i + \alpha h, \tilde{y}_i + \beta hf(x_i, \tilde{y}_i))|$$

$$= |y_{i+1} - \tilde{y}_i - (1-s)hf(x_i, \tilde{y}_i) + (1-s)hf(x_i, y_i) - (1-s)hf(x_i, y_i)$$

$$- shf\left(x_i + \frac{h}{25}, \tilde{y}_i + \frac{h}{2s}f(x_i, \tilde{y}_i)\right) + shf\left(x_i + \frac{h}{2s}, y_i + \frac{h}{2s}f(x_i, y_i)\right)$$

$$- shf\left(x_i + \frac{h}{2s}, y_i + \frac{h}{2s}f(x_i, y_i)\right)|$$

$$\leqslant K|1 - s|h|y_i - \tilde{y}_i| + K|s|h\left|\tilde{y}_i + \frac{h}{2s}f(x_i, \tilde{y}_i) - y_i - \frac{h}{2s}f(x_i, y_i)\right|$$

$$+ \left|y_{i+1} - \tilde{y}_i - y_i + y_i - (1-s)hy_i' - shf\left(x_i + \frac{h}{2s}, y_i + \frac{h}{2s}f(x_i, y_i)\right)\right|$$

$$\leqslant K|1 - s|h|y_i - \tilde{y}_i| + K|s|h|y_i - \tilde{y}_i| + \frac{K^2h^2}{2}|y_i - \tilde{y}_i|$$

$$+ |y_i - \tilde{y}_i| + \left|y_{i+1} - y_i - (1-s)hy_i' - shf\left(x_i + \frac{h}{2s}, y_i + \frac{h}{2s}f(x_i, y_i)\right)\right|$$

$$\leqslant c_1|y_i - \tilde{y}_i| + c_2 \tag{6.8}$$

where

$$c_1 = \left(1 + Kh(|1 - s| + |s|) + \frac{K^2h^2}{2}\right)$$

$$c_2 = \left|y_{i+1} - y_i - (1-s)hy_i' - shf\left(x_i + \frac{h}{2s}, y_i + \frac{h}{2s}f(x_i, y_i)\right)\right|$$

Let us estimate c_2. Consecutively we obtain

$$c_2 = \left| y_{i+1} - y_i - (1-s)hy_i' - shf\left(x_i + \frac{h}{2s}, y_i + \frac{h}{2s}f(x_i, y_i)\right) - hy_{i+1/2}' + hy_{i+1/2}' \right.$$
$$\left. - shf\left(x_i + \frac{h}{2s}, y_{i+1/2s}\right) + shf\left(x_i + \frac{h}{2s}, y_{i+1/2s}\right) \right|$$
$$\leq |y_{i+1} - y_i - hy_{i+1/2}'| + h|y_{i+1/2}' - (1-s)y_i' - sy_{i+1/2s}'|$$
$$+ |s|h \left| f(x_{i+1/2s}, y_{i+1/2s}) - f\left(x_{i+1/2s}, y_i + \frac{h}{2s}f(x_i, y_i)\right) \right| \qquad (6.9)$$

where

$$x_{i+1/2s} = x_i + \frac{h}{2s}, \quad y_{i+1/2s} = y(x_{i+1/2s}), \quad y_{i+1/2s}' = y'(x_{i+1/2s})$$

Let us estimate the first term in the right-hand side of (6.9):

$$|y_{i+1} - y_i - hy_{i+1/2}'| = h\left| \frac{y_{i+1} - y_i}{h} - y_{i+1/2}' \right|$$
$$= \left| \int_{x_i}^{x_{i+1}} [y'(t) - y_{i+1/2}'] \, dt \right|$$
$$= h\left| \int_{-1/2}^{1/2} [y'(x_{i+1/2} + th) - y_{i+1/2}'] \, dt \right|$$
$$= h\left| \int_{0}^{1/2} [y'(x_{i+1/2} + th) - 2y_{i+1/2}' + y'(x_{i+1/2} - th)] \, dt \right|$$
$$\leq h \int_{0}^{1/2} \omega_2\left(y', x_{i+1/2}, \frac{h}{2}\right) dt$$
$$= \frac{h}{2} \omega_2\left(y', x_{i+1/2}; \frac{h}{2}\right). \qquad (6.10)$$

On the other hand

$$\left| f(x_{i+1/2s}, y_{i+1/2s}) - f\left(x_{i+1/2s}, y_i + \frac{h}{2s}f(x_i, y_i)\right) \right|$$
$$\leq K \left| y_{i+1/2s} - y_i - \frac{h}{2s}y_i' \right|$$
$$= \frac{Kh}{2|s|} \left| \frac{y_{i+1/2s} - y_i}{h/2s} - y_i' \right|$$
$$= \frac{Kh}{2|s|} \left| \frac{2s}{h} \int_{x_i}^{x_{i+1/2s}} [y'(t) - y_i'] \, dt \right|$$

$$\leqslant \frac{Kh}{2|s|}\omega\left(y',x_{i+1/4s};\frac{h}{2|s|}\right)$$

$$\leqslant \frac{Kh}{2|s|}\omega\left(y',x_{i+1/2};\left(1+\frac{1}{|s|}\right)h\right) \tag{6.11}$$

It remains to estimate the term $|y'_{i+1/2}-(1-s)y'_i-sy'_{i+1/2s}|$. To this end we use the notation

$$a = \min(x_i,x_{i+1/2},x_{i+1/2s}) \tag{6.12}$$
$$b = \max(x_i,x_{i+1/2},x_{i+1/2s})$$

Let P be the algebraic polynomial of first degree which interpolates the function y' at the points a and b. From lemma 2.3 we have the estimate

$$\|y'-P\|_{C[a,b]}\leqslant\omega_2\left(y',\frac{a+b}{2};\frac{b-a}{2}\right) \tag{6.13}$$

with the usual notation

$$P(x_i)=P_i,\quad P(x_{i+1/2s})=P_{i+1/2s}.$$

it follows from (6.12) and (6.13) that

$$|y'_{i+1/2}-(1-s)y'_i-sy'_{i+1/2s}|\leqslant|y'_{i+1/2}-P_{i+1/2}-(1-s)(y'_i-P_i)$$
$$-s(y'_{i+1/2s}-P_{i+1/2s})|+|P_{i+1/2}-(1-s)P_i-sP_{i+1/2s}|$$
$$\leqslant|y'_{i+1/2}-P_{i+1/2}|+|1-s||y'_i-P_i|+|s||y'_{i+1/2s}-P_{i+1/2s}|$$
$$\leqslant(1+|1-s|+|s|)\omega_2\left(y',\frac{a+b}{2};\frac{b-a}{2}\right) \tag{6.14}$$

since

$$P_{i+1/2}-(1-s)P_i-sP_{i+1/2s}=0$$

if P is an algebraic polynomial of first degree.

From (6.12) it follows that there exist constants $c_3=c_3(s)\geqslant0$ and $c_4=c_4(s)$, such that

$$\frac{b-a}{2}=c_3(s)h,\quad \frac{a+b}{2}=x_{i+1/2}+c_4(s)h \tag{6.15}$$

From (6.14) and (6.15) we obtain

$$|y'_{i+1/2}-(1-s)y'_i-sy'_{i+1/2s}|\leqslant(1+|s|+|1-s|)\omega_2(y',x_{i+1/2}+c_4(s)h;c_3(s)h)$$
$$\leqslant(1+|s|+|1-s|)\omega_2(y',x_{i+1/2};c_5(s)h) \tag{6.16}$$

where $c_5(s)=|c_4(s)|+c_3(s)$.

From (6.8), (6.10), (6.9), (6.11) and (6.16) we obtain

$$|y_{i+1}-\tilde{y}_{i+1}|\leqslant\left[1+Kh\left(|1-s|+|s|+\frac{Kh}{2}\right)\right]|y_i-\tilde{y}_i|+\frac{h}{2}\omega_2\left(y',x_{i+1/2};\frac{h}{2}\right)$$

$$+ \frac{Kh^2}{2} \omega\left(y', x_{i+1/2}; \left(1 + \frac{1}{|s|}\right)h\right)$$

$$+ (1 + |s| + |1 - s|)h\omega_2(y', x_{i+1/2}; c_5(s)h)$$

Recursively applying the above inequality, we obtain

$$|y_{i+1} - \tilde{y}_{i+1}| \leq \sum_{k=0}^{i}\left[1 + Kh(|1-s| + |s| + \frac{Kh}{2}\right]^{i-k}\left[\frac{h}{2}\omega_2\left(y', x_{k+1/2}; \frac{h}{2}\right)\right.$$

$$+ \frac{Kh^2}{2}\omega\left(y', x_{k+1/2}; \left(1 + \frac{1}{|s|}\right)h\right)$$

$$\left. + (1 + |s| + |1-s|)h\omega_2(y', x_{k+1/2}; c_5(s)h)\right] \tag{6.17}$$

We use the notation

$$|s| + |1-s| + \frac{Kh}{2} = |s| + |1-s| + \frac{KA}{2n} \leq |s| + |1-s| + \frac{KA}{2} = c_6(s)$$

$$1 + |s| + |1-s| = c_7(s)$$

Then from (6.17) we obtain

$$\max\{|y_i - \tilde{y}_i| : 0 \leq i \leq n\} \leq \left(1 + \frac{c_6(s)KA}{n}\right)^n \sum_{k=0}^{n-1}\left[\frac{h}{2}\omega_2\left(y', x_{k+1/2}; \frac{h}{2}\right)\right.$$

$$+ \frac{Kh^2}{2}\omega\left(y', x_{k+1/2}; \left(1 + \frac{1}{|s|}\right)h\right)$$

$$\left. + c_7(s)h\omega_2(y', x_{k+1/2}; c_5(s)h)\right]$$

$$\leq e^{c_6(s)KA} \sum_{k=0}^{n-1}\left[\frac{1}{2}\int_{x_k}^{x_{k+1}} \omega_2\left(y', x_{k+1/2}; \frac{h}{2}\right)dx\right.$$

$$+ \frac{Kh}{2}\int_{x_k}^{x_{k+1}}\omega\left(y', x_{k+1/2}; \left(1 + \frac{1}{|s|}\right)h\right)dx$$

$$\left. + c_7(s)\int_{x_k}^{x_{k+1}}\omega_2(y', x_{k+1/2}; c_5(s)h)\,dx\right]$$

$$\leq e^{c_6(s)KA} \sum_{k=0}^{n-1}\left[\frac{1}{2}\int_{x_k}^{x_{k+1}} \omega_2(y', x; h)\,dx\right.$$

$$+ \frac{Kh}{2}\int_{x_k}^{x_{k+1}}\omega\left(y', x; \left(3 + \frac{1}{|s|}\right)h\right)dx$$

$$\left. + c_7(s)\int_{x_k}^{x_{k+1}}\omega_2(y', x; (c_5(s) + 1)h)\,dx\right]$$

$$= e^{c_6(s)KA} \int_0^A \left[\tfrac{1}{2}\omega_2(y', x; h) + \frac{Kh}{2}\omega\left(y', x; \left(3 + \frac{1}{|s|}\right)h \right) \right.$$

$$\left. + c_7(s)\omega_2(y', x; (c_5(s) + 1)h) \right] dx$$

$$= e^{c_6(s)KA} A \left[\tfrac{1}{2}\tau_2(y'; h)_{L[0,A]} \right.$$

$$+ \frac{Kh}{2}\tau\left(y'; \left(3 + \frac{1}{|s|}\right)h \right)_{L[0,A]}$$

$$\left. + c_7(s)\tau_2(y'; (c_5(s) + 1)h)_{L[0,A]} \right]$$

From the last inequality, using property (5) of the modulus $\tau_2(f; \delta)_L$, we obtain the following.

Theorem 6.2. *For the numerical solution of problem* (6.1) *under condition* (6.2) *using the method of* (6.5) *and* (6.6) *the following estimate holds*:

$$\max\{|y_i - \tilde{y}_i|: 0 \leqslant i \leqslant n\} \leqslant c[\tau_2(y'; h)_{L[0,A]} + h\tau(y'; h)_{L[0,A]}]$$

where c is a constant, dependent only on s, K and A.

Remark. From the proof of Theorem 6.2 it follows that for every concrete choice of the parameter s the constant c can be estimated effectively.

From Theorems 6.1 and 6.2 and the properties of the averaged moduli of smoothness we can obtain many consequences which we shall not present here. We wish to remark only that the classical orders of the error $O(h)$ and $O(h^2)$ are obtained under weaker assumptions on the solution y – for the order $O(h)$ in Euler's method it is sufficient for y' to have bounded variation, for the method (6.5) and (6.6) we obtain order $O(h^2)$ if y'' has bounded variation, while there is the usual restriction on y''' to be bounded.

6.2. Adams–Bashforth methods

We shall restrict ourselves only to the extrapolation methods of the Adams–Bashforth type. As is well known, the extrapolation methods of the Adams–Bashforth type for the numerical solution of problem (6.1) can be obtained in the following way.

Let us assume that we have obtained approximate values of the solution y: $\tilde{y}_0, \tilde{y}_1, \ldots, \tilde{y}_i$, at the points $x_j = jh, j = 0, 1, \ldots, i$. Using these values it is possible to compute approximately $\tilde{y}'_j = f(x_j, \tilde{y}_j), j = 0, 1, \ldots, i$. We obtain an approximation of $\tilde{y}'$ in the interval $[x_i, x_{i+1}]$ by interpolating $\tilde{y}$ at the points x_i, $x_{i-1}, \ldots, x_{i-k}$.

Using the Newton backward interpolation formula we obtain

$$\tilde{y}'(x_i + th) = \tilde{y}'(x_i) + \frac{t}{1!}\Delta_h\tilde{y}'_{i-1} + \frac{t(t+1)}{2!}\Delta_h^2\tilde{y}'_{i-2} + \cdots$$

$$+ \frac{t(t+1)\cdots(t+k-1)}{k!}\Delta_h^k\tilde{y}'_{i-k} + R_k(t) \quad x - x_i = th, \qquad (6.18)$$

where $R_k(t)$ is the error of the interpolation formula. Let us ignore the error $R_k(t)$ and integrate (6.18) with respect to t from $-j$ to 1. We obtain

$$\int_{-j}^1 \tilde{y}'(x_i + th)\,dt = \int_{-j}^1 \left[\tilde{y}'_i + \frac{t}{1!}\Delta_h\tilde{y}'_{i-1} + \cdots + \frac{t(t+1)\cdots(t+k-1)}{k!}\Delta_h^k\tilde{y}'_{i-k} \right] dt$$

or

$$\tilde{y}_{i+1} - \tilde{y}_{i-j} = h \left[(1+j)\tilde{y}'_i + \sum_{\theta=1}^k a_{\theta j}\Delta_h^\theta\tilde{y}'_{i-\theta} \right] \qquad (6.19)$$

where

$$a_{\theta j} = \frac{1}{\theta!} \int_{-j}^1 t(t+1)\cdots(t+\theta-1)\,dt$$

If we replace the finite differences in (6.19) using the formula

$$\Delta_h^s\tilde{y}'_{i-s} = \sum_{m=0}^s (-1)^{s+m}\binom{s}{m}\tilde{y}'_{i-s+m}$$

then (6.19) can be rewritten in the form

$$\tilde{y}_{i+1} = \tilde{y}_{i-j} + h \sum_{m=0}^k \beta_{jm}\tilde{y}'_{i-m} \qquad (6.20)$$

If $\tilde{y}'$ is a polynomial of kth degree, then $R_k(t)$ in the formula (5.18) is zero, and therefore for every algebraic polynomial P of kth degree we have the equation

$$\int_{-j}^1 P(x_i + ht)\,dt = \int_{-j}^1 \left[P_i + \frac{t}{1!}\Delta_h P_{i-1} + \cdots + \frac{t(t+1)\cdots(t+k-1)}{k!}\Delta_h^k P_{i-k} \right] dt$$

Therefore

$$\int_{-j}^1 P(x_i + th)\,dt = \sum_{m=0}^k \beta_{im}P_{i-m} \qquad (6.21)$$

Let us now estimate $\varepsilon_i = y_i - \tilde{y}_i$, $i = 0, 1, \ldots, n$. Again we wish to find the solution in the interval $[x_0, x_0 + A]$. Set $h = A/n$ and assume that the condition (6.2) is satisfied. The exact solution y of (6.1) will not satisfy equations (6.20), but will satisfy equations of the type

$$y_{i+1} = y_{i-j} + h \sum_{m=0}^k \beta_{im}y'_{i-m} + r_i. \qquad (6.22)$$

Let us now estimate $|r_i|$. We consider the linear functional

$$L(g) = \int_{-j}^{1} g(x_i + th)\,dt - \sum_{m=0}^{k} \beta_{jm} g_{i-m}$$

From (6.21) it follows that if P is an algebraic polynomial of kth degree, then $L(P) = 0$. Our aim will be to apply Theorem 2.4 from Chapter 2. To do this we must estimate $\|L\|_{C[a,b]}$, where $[a,b] = [x_{i-j}, x_{i+1}]$. We have

$$|L_g| \leqslant (1+j)\|g\|_{C[a,b]} + \sum_{m=0}^{k} |\beta_{jm}| \|g\|_{C[a,b]} + \left(1 + j + \sum_{m=0}^{k} |\beta_{jm}|\right)\|g\|_{C[a,b]}$$

from which

$$\|L\|_{M[a,b]} \leqslant c(k,j)$$

where the constant $c(k,j)$ is given by

$$c(k,j) = 1 + j + \sum_{m=0}^{k} |\beta_{jm}| \tag{6.23}$$

If we apply Theorem 2.4, taking into account that $L(y') = r_i/h$, we obtain

$$|r_i| \leqslant hc(k,j)W_{k+1}\omega_{k+1}\left(y', x_i; 2(1+j)\frac{h}{k+1}\right) \tag{6.24}$$

where $c(k,j)$ is given by (6.23).

If we subtract (6.20) from equation (6.22), in view of the Lipschitz condition (6.2), we obtain

$$\varepsilon_{i+1} = y_{i+1} - \tilde{y}_{i+1}$$

$$= y_{i-j} - \tilde{y}_{i-j} + h\sum_{m=0}^{k} \beta_{jm}(y'_{i-m} - \tilde{y}'_{i-m}) + r_i$$

$$= \varepsilon_{i-j} + h\sum_{m=0}^{k} \beta_{jm}[f(x_{i-m}, y_{i-m}) - f(x_{i-m}, \tilde{y}_{i-m})] + r_i$$

i.e.

$$|\varepsilon_{i+1}| \leqslant |\varepsilon_{i-j}| + hK\sum_{m=0}^{k} |\beta_{jm}||\varepsilon_{i-m}| + |r_i| \tag{6.25}$$

In what follows we shall assume that $j \leqslant k$ (usually $j = 0$ or $j = 1$). Let us assume that use have found, in some way, the values $\tilde{y}_0(=y_0), \tilde{y}_1, \ldots, \tilde{y}_k$ (for example using the Runge–Kutta method). Let

$$|\varepsilon_i| = |y_i - \tilde{y}_i| \leqslant \delta \quad i = 0, 1, \ldots, k \tag{6.26}$$

Using $\tilde{y}_0, \tilde{y}_1, \ldots, \tilde{y}_k$, we can obtain from (6.20) $\tilde{y}_{k+1}$ and subsequently $\tilde{y}_{k+2}$, $\tilde{y}_{k+3}, \tilde{y}_{k+4}, \ldots, \tilde{y}_n$. For the error, setting

$$\sum_{m=0}^{k} |\beta_{jm}| = B \tag{6.27}$$

we obtain from (6.25)

$$|\varepsilon_{k+1}| \leqslant (1+hKB)\delta + |r_k|$$
$$|\varepsilon_{k+2}| \leqslant (1+hKB)^2\delta + (1+hKB)|r_k| + |r_{k+1}|$$
$$\vdots$$
$$|\varepsilon_n| \leqslant (1+hKB)^{n-k}\delta + \sum_{m=k}^{n-1}(1+kKB)^{n-m-1}|r_m|$$

Hence

$$\max\{|\varepsilon_i|: 0 \leqslant i \leqslant n\} \leqslant \left(1 + \frac{A}{n}KB\right)^n\left(\delta + \sum_{m=k}^{n-1}|r_m|\right)$$

$$\leqslant e^{KAB}\left(\delta + \sum_{m=k}^{n-1}|r_m|\right) \qquad (6.28)$$

For $\sum_{m=k}^{n-1}|r_m|$ we obtain, from (6.24),

$$\sum_{m=k}^{n-1}|r_m| \leqslant c(k,j)W_{k+1}h\sum_{m=k}^{n-1}\omega_{k+1}\left(y', x_m; 2(1+j)\frac{h}{k+1}\right)$$

$$= c'(k,j)\sum_{m=k}^{n-1}\int_{x_m}^{x_{m+1}}\omega_{k+1}\left(y', x_m; 2(1+j)\frac{h}{k+1}\right)dx$$

$$\leqslant c'(k,j)\sum_{m=k}^{n-1}\int_{x_m}^{x_{m+1}}\omega_{k+1}\left(y', x; 2(2+j)\frac{h}{k+1}\right)dx$$

$$\leqslant c'(k,j)\int_0^A\omega_{k+1}\left(y', x; 2(2+j)\frac{h}{k+1}\right)dx$$

$$= Ac'(k,j)\tau_{k+1}\left(y'; 2(2+j)\frac{h}{k+1}\right)_{L[0,A]}$$

$$\leqslant Ac''(k,j)\tau_{k+1}(y'; h)_L \qquad (6.29)$$

where in the last inequality we have used property (5) of the moduli $\tau_{k+1}(f; \delta)_L$, and $c''(k,j)$ is a constant which depends only on k and j.

From (6.28) and (6.29) we obtain

$$\max\{|\varepsilon_i|: 0 \leqslant i \leqslant n\} \leqslant e^{KAB}(\delta + Ac''(k,j)\tau_{k+1}(y'; h)_{L[0,A]})$$

So we have proved the following theorem.

Theorem 6.3. *For the Adams–Bashforth extrapolation method (6.20) for the numerical solution of equation (6.1) using condition (6.2), the following estimate holds:*

$$\max\{|y_i - \tilde{y}_i|: 0 \leqslant i \leqslant n\} \leqslant e^{KAB}[\delta + Ac''(k,j)\tau_{k+1}(y'; h)_{L[0,A]}]$$

where δ and B are given by (6.26) and (6.27), and $c''(k,j)$ is a constant, depending only on k and j.

Remark. The constant $c''(k,j)$ for given k and j can be calculated explicitly if we follow the proof of Theorem 6.3.

From Theorem 6.3 for the concrete Adams–Bashforth extrapolation method it is possible to derive many consequences if we make some assumptions about y'. Without formulating these corollaries here we wish to remark that the corresponding orders $O(h^k)$ can be obtained under weaker assumptions than the well-known classical assumptions for the boundedness of the corresponding derivatives of y.

6.3. Notes

The results of Sections 6.1 and 6.2 are given by Andreev [1], [2], [4] (see also Popov and Andreev [1]).

Andreev and Gichev [1] obtain a generalization of the results of Section 6.2. Using the notation of Sections 6.1 and 6.2 let us consider the numerical solution of problem (6.1) using condition (6.2) and the following difference scheme:

$$\sum_{i=0}^{k} a_i \tilde{y}_{n-i} + h \sum_{i=0}^{k} b_i \tilde{y}'_{n-i} = 0$$

where $\tilde{y}_0, \tilde{y}_1, \ldots, \tilde{y}_{k-1}$ are given, $\varepsilon_i = y_i - \tilde{y}_i$, $|\varepsilon_i| \leqslant \delta$, $i = 0, 1, \ldots, k-1$, $a_0 \neq 0$.
We assume the conditions

$$\sum_{i=0}^{k} a_i = 0$$

$$\sum_{i=0}^{k} i^s b_i = \frac{1}{i+1} \sum_{i=0}^{k} a_i i^{s+1} \quad s = 0, 1, \ldots, r$$

These conditions are equivalent to the condition that the linear functional

$$T(u) = \sum_{i=0}^{k} a_i u_{n-i} + h \sum_{i=0}^{k} b_i u'_{n-i}$$

satisfies $T(u) = 0$ for every $u \in H_{r+1}$. Then

$$\max_{0 \leqslant i \leqslant m} |\varepsilon_i| \leqslant \rho^m B^m e^{KAC/B}[\delta + c\tau_{r+1}(y'; h)_{L_1}]$$

where

$$\rho = \left(1 - hK \left|\frac{b_0}{a_0}\right|\right)^{-1}$$

$$B = \sum_{i=1}^{k} \left|\frac{a_i}{a_0}\right|$$

$$C = \sum_{i=1}^{k} \left|\frac{b_i}{a_0}\right|.$$

For the Milne method

$$\tilde{y}_i^{(1)} = \tilde{y}_{i-4} + \frac{4h}{3}(2\tilde{y}'_{i-3} - \tilde{y}'_{i-2} + 2\tilde{y}'_{i-1})$$

$$\tilde{y}_i = y_{i-4} + \frac{h}{3}(\tilde{y}'_{i-2} + 4\tilde{y}'_{i-1} + \tilde{y}'^{(1)}_i)$$

the following estimate is obtained:

$$\max_{0 \leqslant i \leqslant n} |\varepsilon_i| = O[\delta + \tau_4(y'; h)_{L_1}]$$

In the same paper, Andreev and Gichev consider the problem of stability of the corresponding difference schemes.

Popov and Dechevski [1] consider the problem of estimation of the error of the numerical solution of the Cauchy problem for the parabolic equation in network norms. The error is estimated using the averaged moduli of smoothness of the initial condition.

7 ESTIMATION OF THE ERROR IN THE NUMERICAL SOLUTION OF THE BOUNDARY VALUE PROBLEM FOR ORDINARY DIFFERENTIAL EQUATIONS OF SECOND ORDER

We shall consider here the problem of estimating the error of the numerical solution of the boundary value problem for linear differential equations of second order:

$$[k(x)u'(x)]' - q(x)u(x) = -f(x) \quad x \in [0, 1] \tag{7.1}$$

To illustrate the ideas, we shall consider only the simplest boundary conditions

$$u(0) = \alpha, \quad u(1) = \beta \tag{7.2}$$

As usual we shall assume that $q(x) \geq 0$ for $x \in [0, 1]$ and that $k(x) \geq c_0 > 0$ for $x \in [0, 1]$.

There exist many numerical methods for solving the problem of (7.1) and (7.2). We shall consider here only three of them.

In Section 7.1 we develop the technique we shall use and prove some basic lemmas. In Section 7.2 we obtain a general estimate by means of the averaged moduli for the error of the numerical solution of the problem (7.1), (7.2) by means of isotropic differential schemes. In Section 7.3 we obtain an estimate for the error of the method of collocation splines (quadratic and cubic). We consider, in Section 7.4, the error of the simplest finite element method.

7.1.   Auxiliary results

In what follows all functions considered will be defined on the interval $[0, 1]$. We let $x_i = i/n$, $i = 0, \ldots, n$, $h = 1/n$. For every function g we let $g_i = g(x_i)$, $g_{i-1/2} = g(x_i - h/2)$. Following the notation of Samarskii [1], [2], we write

$$a_{x,i} = \frac{a_{i+1} - a_i}{h}$$

$$a_{\bar{x},i} = \frac{a_i - a_{i-1}}{h}$$

We shall also use index-free notation, for example $a_x = \varphi$ will denote $a_{x,i} = \varphi_i$, $i = 1, \ldots, n-1$ (i.e. for every inner point $x_i, i = 1, \ldots, n-1$).

The following lemmas are basic for our purpose.

Lemma 7.1.   (Samarskii [2]). *Let the following differential scheme be given:*

$$(az_{\bar{x}})_x - dz = -\psi, \quad a(x) \geqslant c_0 > 0$$
$$d(x) \geqslant 0, \quad z(0) = z(1) = 0, \quad \psi = \eta_x + \psi^* \tag{7.3}$$

Then

$$\|z\|_{c,h} = \max\{|z_i| : 0 \leqslant i \leqslant n\} \leqslant \frac{2}{c_0}\left(\sum_{i=1}^{n} h|\eta_i| + \sum_{i=2}^{n} h|\mu_i|\right)$$

where

$$\mu_i = \sum_{k=1}^{i-1} h\psi_k^*$$

The notation of (7.3) means that

$$\frac{a_{i+1} z_{\bar{x},i+1} - a_i z_{\bar{x},i}}{h} - d_i z_i = -\psi_i$$

where $i = 1, \ldots, n-1$, $a_i \geqslant c_0 > 0$, $d_i \geqslant 0, i = 0, \ldots, n$.

Proof.   Obviously it is sufficient to consider the case $\psi^* = 0$, since it is easy to see that $\mu_x = \psi^*$. Then the equation for z has the form

$$(az_{\bar{x}})_x - dz = -\eta_x \tag{7.4}$$

Let us consider first the equation

$$(aw_{\bar{x}})_x = -\eta_x \quad w(0) = w(1) = 0 \tag{7.5}$$

From this it follows that $(aw_{\bar{x}} + \eta)_x = 0$, i.e.

$$aw_{\bar{x}} + \eta = \text{const} = c_1$$

Let us represent w_k recursively by w_{k-1} and η_k from the above equation:

$$w_k = w_{k-1} - \frac{h\eta_k}{a_k} + \frac{hc_1}{a_k} \quad k = 1, 2, \ldots, n$$

If we sum with respect to k from 1 to i, we obtain

$$w_i = w_0 - \sum_{k=1}^{i} \frac{h\eta_k}{a_k} + c_1 A_i \quad A_i = \sum_{k=1}^{i} \frac{h}{a_k}$$

Letting $i = n$ and taking into account the boundary condition $w_n = w_0 = 0$, we obtain the constant c_1:

$$c_1 = \frac{1}{A_n} \sum_{k=1}^{n} \frac{h\eta_k}{a_k}$$

Therefore

$$w_i = -\sum_{k=1}^{i} \frac{h\eta_k}{a_k} + \frac{A_i}{A_n} \sum_{k=1}^{n} \frac{h\eta_k}{a_k}$$

$$= -\left(1 - \frac{A_i}{A_n}\right) \sum_{k=1}^{i} \frac{h\eta_k}{a_k} + \frac{A_i}{A_n} \sum_{k=i+1}^{n} \frac{h\eta_k}{a_k} \tag{7.6}$$

From this, since $A_i > 0$, $0 < A_i/A_n < 1$, we obtain

$$|w_i| \leqslant \left(1 - \frac{A_i}{A_n}\right) \sum_{k=1}^{i} \frac{h|\eta_k|}{a_k} + \frac{A_i}{A_n} \sum_{k=i+1}^{n} \frac{h|\eta_k|}{a_k}$$

$$\leqslant \sum_{k=1}^{n} \frac{h|\eta_k|}{a_k} \leqslant \frac{1}{c_0} \sum_{k=1}^{n} h|\eta_k| \tag{7.7}$$

Remark. If we have $\eta_i = (\theta_k)_i$ for some θ, then from (7.6) it follows immediately that in the case when a is a constant we can estimate w_i by $\max |\theta_i|$.

From (7.4) and (7.5) it follows that $\xi_i = z_i - w_i$ satisfy the equations

$$(a\xi_{\bar{x}})_x - d\xi = dw \quad \xi_0 = \xi_n = 0$$

Then from Lemma 7.4 (see below) it follows that

$$\|\xi\|_{c,h} \leqslant \|w\|_{c,h} \tag{7.8}$$

From this and (7.7) we obtain

$$\|z\|_{c,h} \leqslant \|z - w\|_{c,h} + \|w\|_{c,h}$$

$$\leqslant 2\|w\|_{c,h}$$

$$\leqslant \frac{2}{c_0} \sum_{k=1}^{n} h|\eta_k| \qquad \blacksquare$$

It remains to verify (7.8). We can do this using the maximal principle for a tri-diagonal system of linear algebraic equations.

Given the system of linear algebraic equations

$$\mathcal{L}(y_i) \equiv A_i y_{i-1} - C_i y_i + B_i y_{i+1} = -F_i \tag{7.9}$$

where $i = 1, \ldots, n - 1$, $y_0 = \mu_1$, $y_n = \mu_2$, the following conditions are satisfied:

$$A_i > 0, \quad B_i > 0, \quad D_i = C_i - A_i - B_i \geqslant 0 \quad i = 1, 2, \ldots, n - 1 \tag{7.10}$$

Lemma 7.2 (The maximum principle). *Let* $\mathscr{L}(y_i) \geqslant 0$, $(\mathscr{L}(y_i) \leqslant 0)$, $i = 1, \ldots,$
$n - 1$, *and* y_i *be a non-constant function (i.e. $y_i \neq y_j$ at least for two indexes). Then*
y_i *cannot attain its largest positive value (smallest negative value) at an inner point*
of the interval (i.e. for index i between 1 and $n - 1$).

   Proof. Let us assume that y_i attains its positive maximum for i_*, $1 \leqslant i_* \leqslant$
$n - 1$, $y_{i_*} = M$. Since y_i is not constant, then we have $i_0 \neq 0, n$ (which can in
particular also be i_*) for which $y_{i_0} = M$, but either $y_{i_0 - 1} < M$, or $y_{i_0 + 1} < M$ is
valid. Then

$$\begin{aligned}
0 &\leqslant A_{i_0} y_{i_0 - 1} - C_{i_0} y_{i_0} + B_{i_0} y_{i_0 + 1} \\
&= A_{i_0} y_{i_0 - 1} - M C_{i_0} + B_{i_0} y_{i_0 + 1} \\
&< A_{i_0} M - C_{i_0} M + B_{i_0} M \\
&= (A_{i_0} + B_{i_0} - C_{i_0}) M
\end{aligned}$$

which contradicts (7.10).
   In the same way we can prove the case $\mathscr{L}(y_i) \leqslant 0$, $i = 1, \ldots, n - 1$. ∎

Corollary 7.1. *If $\mathscr{L}(y_i) \geqslant 0$, $i = 1, 2, \ldots, n - 1$, $y_0 \leqslant 0$, $y_n \leqslant 0$, then $y_i \leqslant 0$, $i = 1, \ldots,$*
$n - 1$.
   If $\mathscr{L}(y_i) \leqslant 0$, $i = 1, \ldots, n - 1$, $y_0 \geqslant 0$, $y_n \geqslant 0$, then $y_i \geqslant 0$, $i = 1, \ldots, n - 1$.

Lemma 7.3 (Comparison lemma). *Let y_i be the solution of problem (7.5), and*
$\bar{y}_i$ *be the solution of the problem*

$$A_i \bar{y}_{i-1} - C_i \bar{y}_i + B_i \bar{y}_{i+1} = -\bar{F}_i \quad i = 1, \ldots, n - 1$$
$$\bar{y}_0 = \bar{\mu}_1, \quad \bar{y}_n = \bar{\mu}_n$$

Let $|F_i| \leqslant \bar{F}_i$, $i = 1, \ldots, n - 1$, $|\mu_1| \leqslant \bar{\mu}_1$, $|\mu_2| \leqslant \bar{\mu}_2$. Then $|y_i| \leqslant \bar{y}_i$ for $i = 0, \ldots, n$.

   Proof. For $\eta_i = \bar{y}_i \pm y_i$ we obtain

$$\mathscr{L}(\eta_i) = -\bar{F}_i \mp F_i \leqslant 0 \quad i = 1, 2, \ldots, n - 1$$
$$\eta_0 = \bar{\mu}_1 \pm \mu_1 \geqslant 0 \quad \eta_n = \bar{\mu}_2 \pm \mu_2 \geqslant 0$$

therefore from Corollary 7.1 we obtain $\eta_i \geqslant 0$, $i = 1, \ldots, n - 1$. ∎

Lemma 7.4. *For the solution of the system (7.9), (7.10) with $\mu_1 = \mu_2 = 0$ and*
$F_i = D_i - \varphi_i$, $i = 1, 2, \ldots, n - 1$, *the following estimate holds:*

$$\|y\|_C \leqslant \|\varphi\|_C, \quad \|g\|_C = \max\{|g_i| : 0 \leqslant i \leqslant n\}$$

   Proof. If $D_i = 0$ for every $i = 1, \ldots, n - 1$, then by Corollary 7.1 we have
$y_i = 0$, $i = 0, \ldots, n$. Now let $D_i \neq 0$, i.e. there exists $D_{i_0} > 0$ $(D_i \geqslant 0$ by the conditions
for every i). Let Y_i be a solution of the problem

$$\mathscr{L}(Y_i) = -D_i|\varphi_i| \quad Y_0 = Y_n, i = 1, \ldots, n - 1$$

By Lemma 7.3 we have $\|y\|_C \leqslant \|Y\|_C$. According to Lemma 7.2 we have

$$Y_{i_0} = \max\{Y_i : 0 < i < n\} \geqslant 0$$

Then

$$B_{i_0}(Y_{i_0+1} - Y_{i_0}) \leqslant 0, \quad A_{i_0}(Y_{i_0} - Y_{i_0-1}) \geqslant 0 \tag{7.11}$$

and from

$$\mathcal{L}(Y_{i_0}) = -D_{i_0}|\varphi_{i_0}| \tag{7.12}$$

it follows that

$$D_{i_0} Y_{i_0} \leqslant D_{i_0}|\varphi_{i_0}| \leqslant D_{i_0}\|\varphi\|_C$$

If $D_{i_0} > 0$, then $Y_{i_0} \leqslant \|\varphi\|_C$ and we have the required estimate. If $D_{i_0} = 0$, then by (7.12) we obtain

$$B_{i_0} Y_{i_0+1} - C_{i_0} Y_{i_0} + A_{i_0} Y_{i_0-1} = -A_{i_0}(Y_{i_0} - Y_{i_0-1}) + B_{i_0}(Y_{i_0+1} - Y_{i_0}) = 0$$

which by (7.11) is possible only when $Y_{i_0+1} = Y_{i_0} = Y_{i_0-1}$, i.e. then the maximal value of Y_i is attained also at the neighbourhood points. Continuing in this way we find that the maximum of Y_i is attained in the point at which $D_i > 0$. ∎

Lemma 7.5. *For the solution of the problem* (7.9), (7.10) *with* $y_0 = y_n = 0$ *we have*

$$\|y\|_C \leqslant 2\|\mathring{y}\|_C$$

where $\mathring{y}$ is the solution of the system

$$B_i(\mathring{y}_{i+1} - \mathring{y}_i) - A_i(\mathring{y}_i - \mathring{y}_{i-1}) = -F_i \quad i = 1, \ldots, n-1$$
$$\mathring{y}_0 = \mathring{y}_n = 0$$

Proof. In fact, the difference $u_i = y_i - \mathring{y}_i$ satisfies

$$\mathcal{L}(u_i) = -D_i \mathring{y}_i \quad u_0 = u_n = 0$$

By Lemma 7.4 we have $\|u\|_C \leqslant \|\mathring{y}\|_C$, whence

$$\|y\|_C \leqslant \|y - \mathring{y}\|_C + \|\mathring{y}\|_C = \|u\|_C + \|\mathring{y}\|_C \leqslant 2\|\mathring{y}\|_C \qquad \blacksquare$$

Lemma 7.6. *Let us consider the system*

$$(py_{\bar{x}})_{x,i} - (y_{i-1}q_{i,i-1} + q_{i,i}y_i + q_{i,i+1}y_{i+1}) = f_i \tag{7.13}$$

where $i = 1, 2, \ldots, n-1$, $y_0 = y_n = 0$, $p \geqslant c_0 > 0$, $q_{i,j} > 0$. *If*

$$\frac{c_0}{h^2} > 2 \max_{i,j} q_{i,j} \tag{7.14}$$

then

$$\|y\|_C \leqslant \frac{4}{c_0} \sum_{i=1}^{n} |\mu_i| \quad \mu_i = \sum_{k=1}^{i-1} h f_i$$

Proof. We write (7.13) in the form

$$\left(\frac{p_{i+1}}{h^2} - q_{i,i+1}\right)(y_{i+1} - y_i) - \left(\frac{p_i}{h^2} - q_{i,i-1}\right)(y_i - y_{i-1})$$
$$- (q_{i,i-1} + q_{i,i} + q_{i,i+1})y_i = f_i \quad i = 1, \ldots, n-1 \tag{7.15}$$
$$y_0 = y_n = 0$$

From condition (7.14) for the system (7.15) we have

$$A_i > 0, \quad B_i > 0, \quad C_i - A_i - B_i \geqslant 0 \quad i = 1, \ldots, n-1$$

Therefore we can apply Lemmas 7.5 and 7.1 to this system, thus giving the required estimate. ∎

Lemma 7.7. *In the notation of Lemma 7.1 we have*

$$|\eta_i| \leqslant c_1 \omega_k(\varphi, \xi_i; h) + |\theta_i| \quad \xi_i \in [x_{i-1}, x_i]$$
$$|\psi_i^*| \leqslant c_2 \omega_k(g, \zeta_i; h) + |\delta_i| \quad \zeta_i \in [x_{i-1}, x_i]$$

Then the following inequality holds:

$$\|z\|_{C,h} \leqslant \frac{2}{c_0} \left[c_1 \tau_k \left(\varphi; \frac{k+2}{k} h \right)_L + c_2 \tau_k \left(g; \frac{k+2}{k} h \right)_L + h \sum_{i=1}^{n} (|\theta_i| + |\delta_i|) \right]$$

Proof. From Lemma 7.1 we obtain

$$\|z\|_{C,h} \leqslant \frac{2}{c_0} \left\{ \sum_{i=1}^{n} h[c_1 \omega_k(\varphi, \xi_i; h) + |\theta_i|] + \sum_{i=1}^{n} h[c_2 \omega_k(g, \xi_i, h) + |\delta_i|] \right\}$$

$$\leqslant \frac{2}{c_0} \left[c_1 \sum_{i=1}^{n} \int_{x_{i-1}}^{x_i} \omega_k \left(\varphi, x; \frac{k+2}{k} h \right) dx \right.$$

$$+ c_2 \sum_{i=1}^{n} \int_{x_{i-1}}^{x_i} \omega_k \left(g, x; \frac{k+2}{k} h \right) dx + h \sum_{i=1}^{n} (|\theta_i| + |\delta_i|) \right]$$

$$= \frac{2}{c_0} \left[c_1 \tau_k \left(\varphi; \frac{k+1}{k} h \right)_L + c_2 \tau_k \left(g; \frac{k+2}{k} \right)_L + h \sum_{i=1}^{n} (|\theta_i| + |\delta_i|) \right]$$

since for $\xi_i, x \in [x_{i-1}, x_i]$ we have $\omega_k(\varphi, \xi_i; h) \leqslant \omega_k(\varphi, x; (k+2/k)h)$ and the same for $\omega_k(g, \xi_i; h)$. ∎

The following well-known lemma will often be used:

Lemma 7.8. *Let the function y have a bounded second derivative y″ in the interval* [0, 1]. *Then*

$$\left| y_i'' - \frac{\Delta_h^2 y_{i-1}}{h^2} \right| \leqslant \tfrac{1}{2} \omega_2(y'', x_i; h) \quad x_i = \frac{i}{n}, i = 1, 2, \ldots, n-1, h = \frac{1}{n}$$

Proof. Using Taylor's formula with integral remainder, applied to y_{i-1} and y_{i+1}, we obtain

$$y_{i+1} = y_i + h y_i' + \int_{x_i}^{x_{i+1}} (t - x_i) y''(t) \, dt$$

$$y_{i-1} = y_i - h y_i' + \int_{x_{i-1}}^{x_i} (x_i - t) y''(t) \, dt$$

Hence we obtain

$$y_i'' - \frac{y_{i+1} - 2y_i + y_{i-1}}{h^2} = \int_{x_{i-1}}^{x_{i+1}} \left[\frac{h/2}{h^2} y_i'' + \frac{|x_i - t|}{h^2} y''(t) \right] dt$$

$$= \int_0^h \frac{h-u}{h^2} [-y''(x_i + u) + 2y_i'' - y''(x_i - u)] \, du$$

Therefore

$$\left| y_i'' - \frac{\Delta_h^2 y_{i-1}}{h^2} \right| \leqslant \int_0^h \frac{h-u}{h^2} \omega_2(y'', x_i; h) \, du$$

$$= \tfrac{1}{2} \omega_2(y'', x_i; h) \qquad \blacksquare$$

Lemma 7.9. *Let f and y be bounded integrable functions. Then*

$$\left| \frac{1}{2h} \int_{-h}^h [f(x) - f(0)] y(x) \, dx \right| \leqslant |y(0)| \omega_2(f, 0; h) + \omega(y, 0; h) \omega(f, 0; h)$$

Proof. In fact, we have

$$\left| \frac{1}{2h} \int_{-h}^h [f(x) - f(0)] y(x) \, dx \right| \leqslant |y(0)| \frac{1}{2h} \left| \int_{-h}^h [f(x) - f(0)] \, dx \right|$$

$$+ \left| \frac{1}{2h} \int_{-h}^h [f(x) - f(0)] [y(x) - y(0)] \, dx \right|$$

$$\leqslant |y(0)| \omega_2(f, 0; h) + \omega(f, 0; h) \omega(y, 0; h). \qquad \blacksquare$$

7.2. Isotropic conservative difference schemes

We shall now solve the problem of (7.1) and (7.2) by the method of isotropic conservative difference schemes (see Samarskii [1], [2]).

We shall assume that the functions k, q and f are functions of bounded variation in the interval $[0, 1]$, and therefore k, q and f can have points of discontinuity only of the first kind. At the points of discontinuity of these functions we use the so-called conjunction conditions, i.e. we require the function u and the flow ku' to be continuous, so that

$$[u] = u(x + 0) - u(x - 0) = 0$$
$$[ku'] = k(x + 0)u'(x + 0) - k(x - 0)u'(x - 0) = 0$$

The estimates which will be obtained here using the averaged moduli of smoothness can also be obtained under weaker assumptions, but for simplicity we shall restrict ourselves only to this case.

Let us solve the problem (7.1), (7.2) numerically using the following difference scheme:

$$(ay_{\bar{x}})_x - dy = -\varphi$$
$$y_0 = \alpha, \quad y_n = \beta \tag{7.16}$$

where a, d, φ and y are network functions, defined on the net $x_i = i/n$, $i = 0, 1, \ldots, n$. We assume that a_i, d_i and φ_i are defined by the following equations (conditions for isotropy of the scheme):

$$a_i = A[k(x_i + sh)]$$
$$d_i = F[q(x_i + sh)] \tag{7.17}$$
$$\varphi_i = F[f(x_i + sh)]$$

where A and F are linear positive functionals acting on the coefficients k and q and the functions f on the right-hand side of equation (7.1) (with respect to the variable s).

We also assume that the linear positive functionals A and F satisfy the following conditions:

(a) The functional $A(p(s))$ depends only on the values of the function p in the interval $[-\frac{1}{2}, \frac{1}{2}]$.
(b) The functional $F(p(s))$ depends only on the values of the function p in the interval $[-\frac{1}{2}, \frac{1}{2}]$.
(c)
$$A(1) = 1, \quad A(s) = -\frac{1}{2}$$
$$F(1) = 1, \quad F(s) = 0,$$

These conditions are natural, because almost all linear positive functionals, which are used in practice, satisfy these conditions. For example we can set

$$a_i A[k(x_i + sh)] = k_{i-1/2} = k\left(x_i - \frac{h}{2}\right)$$

or

$$a_i = \frac{k_{i-1} + k_i}{2}, \quad a_i = \int_{-1}^{0} k(x_i + sh)\, ds$$

$$\varphi_i = \int_{-1/2}^{1/2} f(x_i + sh)\, ds, \quad \varphi_i = f_i$$

$$d_i = q_i, \quad d_i = \int_{-1/2}^{1/2} q(x_i + sh)\, ds$$

These are some of the most often used functionals A and F.

We shall write an equation for the error $z = y - u$ between the solution u of the problem (7.1), (7.2) and the solution of the problem (7.16). For z (as a network function) we have the difference scheme:

$$(az_{\bar{x}})_x - dz = -\psi$$
$$z_0 = z_n = 0 \tag{7.18}$$

where
$$\psi = (au_{\bar{x}})_x - du + \varphi \qquad (7.19)$$

The right-hand side (following Samarskii [1], [2]) can be represented in the form

$$\psi = \eta_x + \psi^*$$

where

$$\eta_i = (au_{\bar{x}})_i - (ku')_{i-1/2}$$

$$\psi_i^* = \varphi_i - \int_{-1/2}^{1/2} f(x_i + sh)\,ds - d_i u_i + \int_{-1/2}^{1/2} q(x_i + sh)u(x_i + sh)\,ds \qquad (7.20)$$

Indeed, if we integrate equation (7.1) from $x_i - h/2$ to $x_i + h/2$, we obtain

$$k_{i+1/2}u'_{i+1/2} - k_{i-1/2}u'_{i-1/2}$$
$$= \int_{-1/2}^{1/2} q(x_i + sh)u\cdot(x_i + sh)\,ds + \int_{-1/2}^{1/2} f(x_i + sh)\,ds \qquad (7.21)$$

From (7.19) and (7.21) we obtain, for $i = 1, 2, \ldots, n-1$,

$$\psi_i = (au_{\bar{x}})_{x,i} - d_i u_i + \varphi_i$$

$$= (au_{\bar{x}})_{i+1} - (ku')_{i+1/2} - (au_{\bar{x}})_i + (ku')_{i-1/2} + \varphi_i - \int_{-1/2}^{1/2} f(x_i + sh)\,ds$$

$$+ d_i u_i - \int_{-1/2}^{1/2} q(x_i + sh)u(x_i + sh)\,ds$$

$$= \eta_{x,i} + \psi_i^*$$

where η_i and ψ_i^* have the form of (7.20).

Therefore, in view of Lemmas 7.1 and 7.7, in order to estimate $\|z\|_{C,h}$ it is sufficient to estimate $|\eta_i|$ and $|\psi_i^*|$.

Let $P(x_i + sh) = k_{i-1} + (s+1)(k_i - k_{i-1})$, $-1 \leqslant s \leqslant 0$, be the algebraic polynomial of first degree which interpolates $k(x_i + sh)$ in the points -1 and 0, i.e. $P(x_{i-1}) = k_{i-1}$, $P(x_i) = k_i$. Then

$$a_i = A[k(x_i + sh)]$$
$$= A[k(x_i + sh) - P(x_i + sh)] + A[P(x_i + sh)] \qquad (7.22)$$

In view of condition (b) for the linear functional A, and the explicit representation of P given above, we obtain

$$A[P(x_i + sh)] = A[k_{i-1} + (s+1)(k_i - k_{i-1})]$$
$$= k_{i-1}A(1) - (k_i - k_{i-1})A(s+1)$$
$$= \frac{k_{i-1} + k_i}{2} \qquad (7.23)$$

Since A is a linear positive functional and $A(c) = c$ for c constant, then

$$|A(k(x_i + sh) - P(x_i + sh)]| \leqslant A(\max\{|k(x_i + sh) - P(x_i + sh)|: -1 \leqslant s \leqslant 0\})$$
$$= \max\{|k(x_i + sh) - P(x_i + sh)|: -1 \leqslant s \leqslant 0\} \quad (7.24)$$

Using lemma 2.3 we obtain

$$\max\{|k(x_i + sh) - P(x_i + sh)|: -1 \leqslant s \leqslant 0\} \leqslant \omega_2\left(k, x_{i-1/2}; \frac{h}{2}\right) \quad (7.25)$$

Since

$$\left|k_{i-1/2} - \frac{k_{i-1} + k_i}{2}\right| \leqslant \tfrac{1}{2}\omega_2\left(k, x_{i-1/2}; \frac{h}{2}\right)$$

$$\left|\frac{u_i - u_{i-1}}{h} - u'_{i-1/2}\right| \leqslant \tfrac{1}{2}\omega_2\left(u', x_{i-1/2}; \frac{h}{2}\right)$$

we obtain from the inequalities (7.22)–(7.25) the following estimate for $|\eta_i|$:

$$|\eta_i| = \left|a_i \frac{u_i - u_{i-1}}{h} - k_{i-1/2}u'_{i-1/2} + a_iu'_{i-1/2} - a_iu'_{i-1/2}\right|$$

$$\leqslant |u'_{i-1/2}||k_{i-1/2} - a_i| + |a_i|\left|u'_{i-1/2} - \frac{u_i - u_{i-1}}{h}\right]$$

$$\leqslant \|u'\|_C\left[\tfrac{1}{2}\omega_2\left(k_1 x_{i-1/2}; \frac{h}{2}\right) + \omega_2\left(k, x_{i-1/2}; \frac{h}{2}\right)\right] + \|k\|_C\omega_2\left(u', x_{i-1/2}; \frac{h}{2}\right)$$

We have found that

$$|\eta_i| \leqslant \tfrac{3}{2}\|u'\|_C\omega_2\left(k, x_{i-1/2}; \frac{h}{2}\right) + \|k\|_C\omega_2\left(u', x_{i-1/2}; \frac{h}{2}\right) \quad (7.26)$$

Let us now estimate $\|\psi_i^*\|$. First we shall estimate

$$\varphi_i - \int_{-1/2}^{1/2} f(x_i + sh)\,ds$$

Let $Q(x_i + sh)$ be the algebraic polynomial of first degree, which interpolates the function f at the points $x_{i-1/2}$ and $x_{i+1/2}$, i.e.

$$Q(x_{i-1/2}) = f(x_{i-1/2}), \quad Q(x_{i+1/2}) = f(x_{i+1/2})$$

The polynomial Q has the form

$$Q(x_i + sh) = f_{i-1/2} + (s + \tfrac{1}{2})(f_{i+1/2} - f_{i-1/2}) \quad (7.27)$$

Again, using lemma 2.3, we obtain

$$\max\{|Q(x_i + sh) - f(x_i + sh)|: -\tfrac{1}{2} \leqslant s \leqslant \tfrac{1}{2}\} \leqslant \omega_2\left(f, x_i; \frac{h}{2}\right) \quad (7.28)$$

Using the fact that F is a linear functional, we obtain

$$\varphi_i = F[f(x_i + sh)]$$
$$= F[f(x_i + sh) - Q(x_i + sh)] + F[Q(x_i + sh)] \quad (7.29)$$

Again, using the fact F is a linear positive functional, property (b) and the explicit form of Q from (7.27), we obtain

$$F[Q(x_i + sh)] = f_{i-1/2}F(1) + (f_{i+1/2} - f_{i-1/2})F(s + \tfrac{1}{2})$$

$$= \frac{(f_{i-1/2} + f_{i+1/2})}{2} \qquad (7.30)$$

Using the facts that $F(1) = 1$ and F is a linear positive functional, from (7.28)–(7.30) we obtain

$$\left| \varphi_i - \int_{-1/2}^{1/2} f(x_i + sh)\,ds \right| \leqslant F[|f(x_i + sh) - Q(x_i + sh)|]$$

$$+ \left| \frac{f_{i-1/2} + f_{i+1/2}}{2} + f_i - f_i - \int_{-1/2}^{1/2} f(x_i + sh)\,ds \right|$$

$$\leqslant \max\{|f(x_i + sh) - Q(x_i + sh)|: -\tfrac{1}{2} \leqslant s \leqslant \tfrac{1}{2}\}$$

$$+ \tfrac{1}{2}|f_{i+1/2} - 2f_i + f_{i-1/2}| + \left| \int_0^{1/2} [f(x_i + sh) - 2f_i + f(x_i - sh)]\,ds \right|$$

$$\leqslant \omega_2\left(f, x_i; \frac{h}{2} \right) + \tfrac{1}{2}\omega_2\left(f, x_i; \frac{h}{2} \right) + \tfrac{1}{2}\omega_2\left(f, x_i; \frac{h}{2} \right)$$

$$= 2\omega_2\left(f, x_i; \frac{h}{2} \right) \qquad (7.31)$$

which is the required estimate.

Let us finally estimate

$$\left| d_i u_i - \int_{-1/2}^{1/2} q(x_i + sh)u(x_i + sh)\,ds \right|$$

Again let $R(x_i + sh)$ be the algebraic polynomial of first degree which interpolates q at the points $x_{i-1/2}$ and $x_{i+1/2}$, i.e.

$$R(x_i + sh) = q_{i-1/2} + (s + \tfrac{1}{2})(q_{i+1} - q_{i-1/2})$$

Using the properties of the linear positive functional F, we obtain the following relations:

$$d_i = F[q(x_i + sh)]$$

$$= F[q(x_i + sh) - R(x_i + sh)] + F[R(x_i + sh)]$$

$$|F[q(x_i + sh) - R(x_i + sh)]| \leqslant \max\{|q(x_i + sh) - R(x_i + sh)|: -\tfrac{1}{2} \leqslant s \leqslant \tfrac{1}{2}\}.$$

$$\leqslant \omega_2\left(q, x_i; \frac{h}{2} \right)$$

$$F[R(x_i + sh)] = F[q_{i-1/2} + (s + \tfrac{1}{2})(q_{i+1/2} - q_{i-1/2})]$$

$$= \frac{q_{i-1/2} + q_{i+1/2}}{2}$$

From these relations we obtain

$$\left| d_i u_i - \int_{-1/2}^{1/2} q(x_i + sh) u(x_i + sh) \, ds \right|$$

$$\leqslant \left| F[q(x_i + sh)] u_i - u_i \int_{-1/2}^{1/2} q(x_i + sh) \, ds \right|$$

$$+ \left| \int_{-1/2}^{1/2} q(x_i + sh)[u_i - u(x_i + sh)] \, ds \right|$$

$$\leqslant \|u\|_c \left\{ \omega_2\left(q, x_i; \frac{h}{2}\right) + \tfrac{1}{2}|q_{i+1/2} - 2q_i + q_{i-1/2}| \right.$$

$$+ \left. \left| \int_{-1/2}^{1/2} [q(x_i + sh) - q_i] \, dx \right| \right\}$$

$$+ \|q\|_c \int_0^{1/2} |u(x_i + sh) - 2u_i + u(x_i - sh)| \, ds$$

$$\leqslant 2\|u\|_c \omega_2\left(q, x_i; \frac{h}{2}\right) + \tfrac{1}{2}\|q\|_c \omega_2\left(u, x_i; \frac{h}{2}\right) \tag{7.32}$$

The inequalities (7.31) and (7.32) give us

$$|\psi_i^*| \leqslant 2\omega_2\left(f, x_i; \frac{h}{2}\right) + 2\|u\|_c \omega_2\left(q, x_i; \frac{h}{2}\right) + \tfrac{1}{2}\|q\|_c \omega_2\left(u, x_i; \frac{h}{2}\right) \tag{7.33}$$

The estimates (7.26) and (7.33) show that for the error $z = y - u$ we can apply Lemma 7.7 or, directly, Lemma 7.1. Let us apply Lemma 7.1. Since A is a positive functional and $k(x) \geqslant c_0 > 0$, we have $a_i = A[k(x_i + sh)] \geqslant A(c_0) = c_0 > 0$ and

$$\|z\|_{C,h} \leqslant \frac{2}{c_0}\left[\tfrac{3}{2}\|u'\|_c \sum_{i=1}^{n} \omega_2\left(k, x_{i-1/2}; \frac{h}{2}\right) h \right.$$

$$+ \tfrac{1}{2}\|k\|_c \sum_{i=1}^{n} \omega_2\left(u', x_{i-1/2}; \frac{h}{2}\right) h + 2\sum_{i=1}^{n-1} \omega_2\left(f, x_i; \frac{h}{2}\right) h$$

$$+ 2\|u\|_c \sum_{i=1}^{n-1} \omega_2\left(q, x_i; \frac{h}{2}\right) h + \tfrac{1}{2}\|q\|_c \sum_{i=1}^{n-1} \omega_2\left(u, x_i; \frac{h}{2}\right) h \left. \right]$$

$$\leqslant \frac{2}{c_0}\left[\tfrac{3}{2}\|u'\|_c \sum_{i=1}^{n} \int_{x_{i-1}}^{x_i} \omega_2(k, x; h) \, dx + \tfrac{1}{2}\|k\|_c \sum_{i=1}^{n} \int_{x_{i-1}}^{x_i} \omega_2(u', x; h) \, dx \right.$$

$$+ 2\sum_{i=1}^{n-1} \int_{x_{i-1/2}}^{x_{i+1/2}} \omega_2(f, x; h) \, dx + 2\|u\|_c \sum_{i=1}^{n-1} \int_{x_{i-1/2}}^{x_{i+1/2}} \omega_2(q, x; h) \, dx$$

$$+ \tfrac{1}{2}\|q\|_c \sum_{i=1}^{n-1} \int_{x_{i-1/2}}^{x_{i+1/2}} \omega_2(u, x; h) \, dx \left. \right]$$

$$\leqslant \frac{2}{c_0}[\tfrac{3}{2}\| u'\|_C \tau_2(k;h)_L$$

$$+ \tfrac{1}{2}\| k\|_C \tau_2(u';h)_L + 2\tau_2(f;h)_L$$

$$+ 2\| u\|_C \tau_2(q;h)_L + \tfrac{1}{2}\| q\|_C \tau_L(u;h)_L]$$

We now formulate the result obtained.

Theorem 7.1. *Let u be the solution of problem (7.1), (7.2) and let y be the solution of problem (7.16), where a, d and φ are given by (7.17), and the linear positive functionals A and F satisfy conditions (a)–(c). Then for the error $z = y - u$ the following estimate holds*

$$\| z\|_{C,h} = \| y - u\|_{C,h}$$

$$= \max \{| y_i - u(x_i) \colon 0 \leqslant i \leqslant n\}$$

$$\leqslant \frac{2}{c_0}[\tfrac{3}{2}\| u'\|_C \tau_2(k;h)_L + \tfrac{1}{2}\| k\|_C \tau_2(y';h)_L$$

$$+ 2\| u\|_C \tau_2(q;h)_L + 2\tau_2(f;h)_L + \tfrac{1}{2}\| q\|_C \tau_2(u;h)_L] \qquad (7.34)$$

We have derived estimate (7.34) under the assumption that the functions k, q and f are of bounded variation, because we do not wish to use the notion of a generalized solution. Indeed, estimate (7.34) is valid under weaker assumptions, for example it is sufficient to assume that k, q and f are Riemann integrable.

Estimate (7.34) looks complicated, but from this we can derive many consequences, if we make some assumptions about the functions k, q and f. For example under the assumption that the functions k, q and f are of bounded variation, it is easy to see that u' also has bounded variation and, using the properties of the averaged moduli, we obtain the following immediately.

Corollary 7.2. *If k, q and f have bounded variation in the interval $[0, 1]$, then*

$$\| y - u\|_{C,h} = O(h)$$

If k, q and f are absolutely continuous functions, then u' is also an absolutely continuous function and from the properties of τ_2 we derive the following corollary.

Corollary 7.3. *If k, q and f are absolutely continuous functions on the interval $[0, 1]$, then*

$$\| y - u\|_{C,h} = O[h\omega(u'';h)_L + h\omega(k';h)_L + h\omega(q';h)_L + h\omega(f';h)_L]$$

From this we obtain the following.

Corollary 7.4. *If the functions k, q and f have derivatives with bounded variation in the interval $[0, 1]$, then $\| y - u\|_{C,h} = O(h^2)$.*

Remark. These corollaries strengthen the estimates which were previously known for the numerical solution of the boundary value problem by method (7.16) (see, for example, Samarskii [1], [2]). The estimate in Corollary 7.4 is obtained there under the assumptions that k, q and f are in $C^2[0, 1]$, i.e. that they have a continuous second derivative in the interval $[0, 1]$ which is stronger than our restriction that k, q and f should have a first derivative with bounded variation. The estimate from Corollary 7.2 is obtained under the assumption that k, q and f are in $Q^2[0, 1]$, i.e. that they have a piecewise continuous second derivative in $[0, 1]$, which is again much stronger than our restriction that the function should have bounded variation.

At the end of this section we shall give an estimate for the error of the so-called 'best scheme'. This scheme has the form

$$(ay_{\bar{x}})_x - dy = -\varphi, \quad y_0 = \alpha, \quad y_n = \beta$$
$$d \geqslant 0, \quad a \geqslant c_0 > 0 \tag{7.35}$$

where the coefficients a, d and φ are given by

$$a_i = \left[\frac{1}{h} \int_{x_{i-1}}^{x_i} \frac{dt}{k(t)} \right]^{-1}$$

$$d_i = \frac{1}{h} \int_{x_{i-1/2}}^{x_{i+1/2}} q(t) \, dt \tag{7.36}$$

$$\varphi_i = \frac{1}{h} \int_{x_{i-1/2}}^{x_{i+1/2}} f(t) \, dt$$

Obviously if $k \geqslant c_0 > 0$, then $a \geqslant c_0 > 0$.

We see that, for the 'best scheme', d and φ are defined by a linear positive functional F which satisfies conditions (a)–(c), but a is given by a functional which is non-linear with respect to k.

For the scheme (7.35), (7.36) we shall apply Lemma 7.1 directly. To this end, we need to estimate

$$\eta_i = a_i u_{\bar{x},i} - k_{i-1/2} u'_{i-1/2}$$

and

$$\psi_i^* = \varphi_i - \int_{-1/2}^{1/2} f(x_i + sh) \, ds - \left[d_i u_i - \int_{-1/2}^{1/2} q(x_i + sh) u(x_i + sh) \, ds \right]$$

Let us first estimate ψ_i^*. From (7.36) we obtain

$$|\psi_i^*| = \left| d_i u_i - \int_{-1/2}^{1/2} q(x_i + sh) u(x_i + sh) \, ds \right|$$

$$\leqslant \left| \int_{-1/2}^{1/2} [q(x_i + sh) - q(x_i)] u_i \, ds \right| + \left| \int_{-1/2}^{1/2} q(x_i) [u(x_i + sh) - u_i] \, ds \right|$$

$$= \left| \int_0^{1/2} u_i [q(x_i + sh) - 2q(x_i) + q(x_i - sh)] \, ds \right|$$

$$+ \left| \int_0^{1/2} q_i [u(x_i + sh) - 2u_i + u(x_i - sh)] \, ds \right|$$

$$\leqslant \|u\|_C \omega_2 \left(q, x_i; \frac{h}{2} \right) + \|q\|_C \omega_2 \left(u, x_i; \frac{h}{2} \right)$$

i.e.

$$|\psi_i^*| \leqslant \|u\|_C \omega_2 \left(q, x_i; \frac{h}{2} \right) + \|q\|_C \omega_2 \left(u, x_i; \frac{h}{2} \right) \tag{7.37}$$

For η_i we have, from (7.36),

$$\eta_i = a_i u_{\bar{x},i} - k_{i-1/2} u'_{i-1/2}$$

$$= h \left(\int_{x_{i-1}}^{x_i} \frac{dt}{k(t)} \right)^{-1} \frac{1}{h} \int_{x_{i-1}}^{x_i} u'(x) \, dx - k_{i-1/2} u'_{i-1/2}$$

$$= \frac{1}{h} \int_{x_{i-1}}^{x_i} [k(x)u'(x) - k_{i-1/2} u'_{i-1/2}] \, dx$$

$$+ \int_{x_{i-1}}^{x_i} \left[\left(\int_{x_{i-1}}^{x_i} \frac{dt}{k(t)} \right)^{-1} - \frac{k(x)}{h} \right] u'(x) \, dx$$

$$= \int_0^{1/2} [k(x_{i-1/2} + sh)u'(x_{i-1/2} + sh) - 2k_{i-1/2} u'_{i-1/2}$$

$$+ k(x_{i-1/2} - sh)u'(x_{i-1/2} - sh)] \, ds$$

$$+ \int_{x_{i-1}}^{x_i} \left[\left(\int_{x_{i-1}}^{x_i} \frac{dt}{k(t)} \right)^{-1} \bigg/ k(x) - \frac{1}{h} \right] [k(x)u'(x) - k_{i-1/2} u'_{i-1/2}] \, dx \tag{7.38}$$

since

$$\int_{x_{i-1}}^{x_i} \left[\left(\int_{x_{i-1}}^{x_i} \frac{dt}{k(t)} \right)^{-1} \bigg/ k(x) - \frac{1}{h} \right] dx = 0$$

We have

$$\int_{x_{i-1}}^{x_i} \frac{dt}{k(t)} = \frac{h}{k(\xi_i)}$$

where $\xi_i \in [x_{i-1}, x_i]$. Therefore

$$\left| \left[k(x) \int_{x_{i-1}}^{x_i} \frac{dt}{k(t)} \right]^{-1} - \frac{1}{h} \right| = \left| \frac{k(\xi_i) - k(x)}{hk(x)} \right| \leqslant \frac{\omega(k, x; h)}{hc_0} \tag{7.39}$$

From (7.38) and (7.39) we obtain:

$$|\eta_i| \leqslant \omega_2 \left(ku', x_{i-1/2}; \frac{h}{2} \right) + \int_{x_{i-1}}^{x_i} \frac{\omega(k, x; h)\omega(ku', x; h/2)}{hc_0} \, dx \tag{7.40}$$

If k, q and f are bounded functions, then $(ku')'$ will also be a bounded function, and

$$\|(ku')'\|_C \leqslant \|f\|_C + \|q\|_C \|u\|_C$$

Therefore

$$\omega\left(ku', x_{i-1/2}; \frac{h}{2}\right) \leqslant Kh \tag{7.41}$$

where

$$K \leqslant \frac{\|f\|_C + \|q\|_C \|u\|_C}{2} \tag{7.42}$$

Therefore (7.40) becomes

$$|\eta_i| \leqslant \omega_2\left(ku', x_{i-1/2}; \frac{h}{2}\right) + \frac{K}{c_0} \int_{x_{i-1}}^{x_i} \omega(k, x; h)\, dx \tag{7.43}$$

Applying Lemma 7.1 and using inequalities (7.37) and (7.43), we obtain the following estimate for the error $z = y - u$ between the solution of the 'best scheme' (7.36) and the solution of the problem u of (7.1), (7.2).

$$\|z\|_{C,h} = \|y - u\|_{C,h} = \max\{|y_i - u_i| : 0 \leqslant i \leqslant n\}$$

$$\leqslant \frac{2}{c_0}\left[\sum_{i=1}^{n} h|\eta_i| + \sum_{i=1}^{n-1} h|\psi_i^*|\right]$$

$$\leqslant \frac{2}{c_0}\left[\sum_{i=1}^{n} h\omega_2\left(ku', x_{i-1}; \frac{h}{2}\right) + \frac{Kh}{c_0}\sum_{i=1}^{n}\int_{x_{i-1}}^{x_i} \omega(k, x; h)\, dx\right.$$

$$= \|u\|_C \sum_{i=1}^{n-1} h\omega_2\left(q, x_i; \frac{h}{2}\right) + \|q\|_C \sum_{i=1}^{n-1} h\omega_2\left(u, x_i; \frac{h}{2}\right)\Big]$$

$$\leqslant \frac{2}{c_0}\left[\sum_{i=1}^{n}\int_{x_{i-1}}^{x_i} \omega_2(ku', x; h)\, dx + \frac{Kh}{c_0}\tau(k; h)_L\right.$$

$$+ \|u\|_C \sum_{i=1}^{n-1}\int_{x_{i-1/2}}^{x_{i+1/2}} \omega_2(q, x; h)\, dx + \|q\|_C \sum_{i=1}^{n-1}\int_{x_{i-1/2}}^{x_{i+1/2}} \omega_2(u, x; h)\, dx\Big]$$

$$= \frac{2}{c_0}\left[\tau_2(ku'; h)_L + \frac{Kh}{c_0}\tau(k; h)_L + \|u\|_C \tau_2(q; h)_L + \|q\|_C \tau_2(n; h)_L\right]$$

This estimate gives us the following theorem for the 'best scheme'.

Theorem 7.2. *For the error $z = y - u$ of the 'best scheme' (7.35), (7.36) for the equation (7.1), (7.2) the following estimate holds:*

$$\|z\|_{C,h} \leqslant \frac{2}{c_0}\left[\tau_2(ku'; h)_L + \|u\|_C \tau_2(q; h)_L + \|q\|_C \tau_2(u; h)_L + \frac{Kh}{c_0}\tau(k; h)_L\right]$$

where the constant K is estimated by (7.42).

Again from the estimate given above we can obtain corollaries which strengthen the corresponding results from Samarskii [1], [2].

Corollary 7.5. *Let the functions k, q and f be functions of bounded variation on* [0, 1]. *Then*

$$\| y - u \|_{C,h} = O(h^2)$$

Indeed, in this case ku' and u' will also be functions of bounded variation and the corollary follows directly from the corresponding property of $\tau(f; \delta)_L$.

Samarskii [1] obtained the estimate $O(h^2)$ under the stronger assumption that the functions k, q and f have piecewise continuous second derivatives.

7.3. Collocation splines

Here we shall consider the problem of estimating the error in the numerical solution of the problem

$$y'' - qy = f \quad x \in [0, 1] \tag{7.44}$$

$$y(0) = \alpha, \quad y(1) = \beta$$

by the method of collocation splines. We shall consider the case of quadratic and cubic splines.

Let us first consider collocation quadratic splines.

As in Section 4.3, we shall represent the quadratic spline by means of the B-splines of Schöenberg:

$$S_2(x) = \sum_{k=-1}^{n+1} c_k B_2(x - kh) \tag{7.45}$$

where

$$B_2(x) = \frac{1}{2h^2} \sum_{j=0}^{3} (-1)^j \binom{3}{j} \left(x + \frac{3h}{2} - jh \right)_+^2, \quad h = \frac{1}{n}$$

is the B_2-spline of second degree.

The coefficients $c_k, k = -1, 0, \ldots, n+1$ will be obtained from the collocation conditions at the points $x_i = i/n, i = 0, 1, \ldots, n$, and from the boundary conditions, i.e. from the equations

$$S_2''(x_i) - q_i S_2(x_i) = f_i \quad i = 0, 1, \ldots, n \tag{7.46}$$

$$S_2(0) = \alpha, \quad S_2(1) = \beta$$

The spline S_2, defined by (7.45) and (7.46), will be the approximation of the solution y of problem (7.44).

From the representation of S_2 and the properties of the B-spline (see Section 4.3) we obtain immediately

$$S_2(x_i) = \frac{c_{i-1} + 6c_i + c_{i+1}}{8} \quad i = 0, 1, \ldots, n$$

$$S_2''(x_i) = \frac{c_{i-1} - 2c_i + c_{i+1}}{h^2} \quad i = 0, 1, \ldots, n$$

$$\tag{7.47}$$

$$S_2''(0) = q_0 S_2(0) + f_0 = y_0''$$
$$S_2''(1) = q_n S_2(1) + f_n = y_n''$$

In view of equations (7.47), equations (7.46) give us the following system of linear algebraic equations for determining the coefficients c_i in the representation of (7.45)

$$c_{-1} + 6c_0 + c_1 = 8y_0$$
$$c_{-1} - 2c_0 + c_1 = h^2 y_0''$$
$$\left(\frac{1}{h^2} - \frac{q_i}{8}\right)c_{i-1} - \left(\frac{2}{h^2} + \frac{3q_i}{4}\right)c_i + \left(\frac{1}{h^2} - \frac{q_i}{8}\right)c_{i+1} = f_i$$

where $i = 1, 2, \ldots, n-1$,

$$c_{n-1} - 2c_n + c_{n+1} = h^2 y_0''$$
$$c_{n-1} + 6c_n + c_{n+1} = 8y_n$$

Subtracting the first two equations we eliminate c_{-1}, and from the last two equations we eliminate c_{n+1}. We obtain the system

$$c_0 = y_0 - \frac{h^2(q_0 y_0 + f_0)}{8}$$

$$\Delta^2 c_{i-1} - \left(\frac{2/h^2 + 3q_i/4}{1/h^2 - q_i/8}\right)c_i = \frac{f_i}{1/h^2 - q_i/8} \quad i = 1, \ldots, n-1 \tag{7.48}$$

$$c_n = y_n - \frac{h^2(q_n y_n + f_n)}{8}$$

Let us define $\theta_i, i = 0, \ldots, n$, by

$$\theta_i = -h^2[\gamma_1 + ih(\gamma_2 - \gamma_1)]$$

where

$$\gamma_1 = \frac{q_0 y_0 + f_0}{8}, \quad \gamma_2 = \frac{q_n y_n + f_n}{8}$$

From (7.48) we find that $r_i = c_i - \theta_i, i = 0, \ldots, n$, satisfy the equations

$$r_0 = y_0$$

$$\Delta^2 r_{i-1} - \frac{q_i r_i}{(1 - h^2 q_i/8)} = \frac{f_i}{1 - h^2 q_i/8} + \frac{q_i \theta_i}{1 - h^2 q_i/8} \quad i = 1, \ldots, n-1 \tag{7.49}$$

$$r_n = y_n$$

From the definition of θ_i it follows, that

$$|\theta_i| = |c_i - r_i| \leqslant h^2 \max(|\gamma_1|, |\gamma_2|) \quad i = 0, \ldots, n \qquad (7.50)$$

The exact solution y of problem (7.44) satisfies the equations

$$y_0 = \alpha$$

$$\frac{\Delta^2 y_{i-1}}{h^2} - q_i y_i = f_i + \varphi_i \quad i = 1, \ldots, n-1$$

$$\varphi_i = \frac{\Delta^2 y_{i-1}}{h^2} - y_i'' \quad i = 1, \ldots, n-1 \qquad (7.51)$$

$$y_n = \beta$$

Let h be so small that $1 - h^2 q_i/8 > \frac{1}{2}$ and $h^2 \|q\|_C/4 < 1$, i.e.

$$\left(1 - \frac{h^2 q_i}{8}\right)^{-1} = 1 + \alpha_i \leqslant 1 + \frac{h^2 \|q\|_C}{4} \leqslant 2 \quad i = 1, \ldots, n-1 \qquad (7.52)$$

or $\alpha_i \leqslant h^2 \|q\|_C/4$, where α_i are defined by (7.52). Then from (7.49), (7.51) and (7.52) it follows that the error $R_i = y_i - z_i$ satisfies the following difference scheme:

$$(R_{\bar{x}})_x - q(1+\alpha)R = \psi \quad x = x_i, i = 1, \ldots, n-1$$

$$R_0 = R_n = 0 \qquad (7.53)$$

where

$$\psi_i = -\alpha_i(q_i y_i + f_i) + q_i(1+\alpha_i)\theta_i + \frac{\Delta^2 y_{i-1}}{h^2} - y_i''$$

Let us now estimate ψ_i. Using (7.50), (7.52) and Lemma 7.8 we obtain

$$|\psi_i| \leqslant \frac{h^2 \|q\|_C}{4}(\|q\|_C \|y\|_C + \|f\|_C)$$

$$+ \left(1 + \frac{h^2}{4}\|q\|_C\right)\|q\|_C h^2 \max\{|\gamma_1|, |\gamma_2|\} + \tfrac{1}{2}\omega_2(y'', x_i; h) \qquad (7.54)$$

Using Lemma 7.1, from (7.53) and (7.54) we get

$$\|R\|_{C,h} = \max\{|y_i - r_i| : 0 \leqslant i \leqslant n\}$$

$$\leqslant \sum_{i=1}^{n} h|y_i|$$

$$\leqslant \tfrac{1}{2} \sum_{i=1}^{n} \int_{x_{i-1/2}}^{x_{i+1/2}} \omega_2(y'', x_i; 2h)\,dx + Kh^2$$

$$\leqslant \tfrac{1}{2}\tau_2(y''; 2h)_L + Kh^2 \qquad (7.55)$$

where the constant K can be estimated by

$$K \leqslant \frac{\|q\|_c}{4}(\|q\|_c\|y\|_c + \|f\|_c) + 2\|q\|_c\left(\frac{\|q\|_c\|y\|_c + \|f\|_c}{8}\right)$$

$$= \|q\|_c\left(\frac{\|q\|_c\|y\|_c + \|f\|_c}{2}\right) \tag{7.56}$$

since

$$|\gamma_1| \leqslant \frac{\|q\|_c\|y\|_c + \|f\|_c}{8}$$

$$|\gamma_2| \leqslant \frac{\|q\|_c\|y\|_c + \|f\|_c}{8}$$

From (7.55) and (7.50) we obtain

$$\max\{|c_i - y_i|: 0 \leqslant i \leqslant n\} \leqslant \max\{|c_i - r_i|: 0 \leqslant i \leqslant n\} + \max\{|y_i - r_i|: 0 \leqslant i \leqslant n\}$$

$$\leqslant h^2\left(\frac{\|q\|_c\|y\|_c + \|f\|_c}{8}\right) + \tfrac{1}{2}\tau_2(y''; 2h)_L + Kh^2$$

$$\leqslant \tfrac{1}{2}\tau_2(y''; 2h)_L + K_1 h^2 \tag{7.57}$$

where (see (7.56))

$$K_1 \leqslant \frac{(\|q\|_c + \tfrac{1}{4})}{2}(\|q\|_c\|y\|_c + \|f\|_c) \tag{7.58}$$

From (7.57) we obtain, in view of the first equation in (7.47) and (7.56), for every $i = 1, \ldots, n-1$,

$$|y_i - S_2(x_i)| = \left|y_i - \frac{c_{i-1} + 6c_i + c_{i+1}}{8}\right|$$

$$\leqslant \tfrac{1}{8}(|y_{i-1} - c_{i-1}| + 6|y_i - c_i| + |y_{i+1} - c_{i+1}|) + \tfrac{1}{8}|\Delta_h^2 y_{i-1}|$$

$$\leqslant \tfrac{1}{2}\tau_2(y''; 2h)_L + K_1 h^2 + \tfrac{1}{2}\omega_2(y; h)_c \tag{7.59}$$

where the constant K_1 can be estimated by (7.58).

On the other hand, from the boundary conditions we have

$$y_0 = S_2(0), \quad y_n = S_2(1) \tag{7.60}$$

Since

$$\omega_2(y; h)_c \leqslant h^2\|y''\|_c \leqslant h^2(\|y\|_c\|q\|_c + \|f\|_c)$$

then (7.59), (7.60) and (7.58) finally give us

$$\max\{|y_i - S_2(x_i): 0 \leqslant i \leqslant n\} \leqslant \tfrac{1}{2}\tau_2(y''; 2h)_L + K_2 h^2 \tag{7.61}$$

where

$$K_2 \leqslant \frac{\|q\|_c + \tfrac{9}{4}}{2}(\|q\|_c\|y\|_c + \|f\|_c) \tag{7.62}$$

From (7.62) and (7.61) we obtain the following.

Theorem 7.3. *For the error in the solution of problem (7.44) under condition (7.52) using the method of collocation quadratic splines (7.45), (7.46), the following estimate holds:*

$$\|y - S_2\|_{C,h} = \max\{|y_i - S_2(x_i)|: 0 \leqslant i \leqslant n\} \leqslant \tfrac{1}{2}\tau_2(y''; 2h)_L + K_2 h^2$$

where the constant K_2 is estimated by (7.62).

From the estimate (7.61) we can also obtain an estimate for

$$\|y - S_2\|_C = \max\{|y(x) - S_2(x)|: x \in [0, 1]\}$$

using Theorem 4.3.

Let $\bar{S}_2$ be the interpolation quadratic spline (see Section 4.3.2) for the solution y of problem (7.44), which interpolates y at the points $x_i = i/n$, $i = 0, 1, \ldots, n$, and satisfies the boundary conditions

$$\bar{S}_2''(0) = \bar{S}_2''(1) = 0$$

Then $S = S_2 - \bar{S}_2$ will be a quadratic spline of type (7.45) and thus by Theorem 4.9 we have

$$\max\{|S(x)|: x \in [0, 1]\} \leqslant 4 \max\{|S(x_i)|: 0 \leqslant i \leqslant n\} + h^2 \max\{|\bar{\alpha}|, |\bar{\beta}|\}$$
$$= 4 \max\{|y_i - S_2(x_i)|: 0 \leqslant i \leqslant n\} + h^2 \max\{|\bar{\alpha}|, |\bar{\beta}|\}$$
$$(7.63)$$

where

$$\bar{\alpha} = S''(0) = S_2''(0) - \bar{S}_2''(0) = S_2''(0)$$
$$\bar{\beta} = S''(1) = S_2''(1) - \bar{S}_2''(1) = S_2''(1)$$

But

$$S_2''(0) = f(0) + q(0)y(0), \quad S_2''(1) = f(1) + q(1)y(1)$$

and therefore

$$\max\{|\bar{\alpha}|, |\bar{\beta}|\} \leqslant \|q\|_C \|y\|_C + \|f\|_C$$

From this, (7.63) and Theorem 7.3 we obtain

$$\max\{|S(x)|: x \in [0, 1]\} \leqslant 2\tau_2(y''; 2h)_L + K_3 h^2 \qquad (7.64)$$

where

$$K_3 \leqslant \left(\frac{\|q\|_C + 17/4}{8}\right)(\|q\|_C \|y\|_C + \|f\|_C)$$

If we look at the proof of Theorem 4.10, it is easy to see that the result can be written in the form

$$\|y - \bar{S}_2\|_C \leqslant 4h^2 \|y''\|_C \qquad (7.65)$$

where $\bar{S}_2$ is the interpolation quadratic spline for y.

From (7.64) and (7.65) it follows that

$$\|S_2 - y\|_C \leqslant \|S_2 - \bar{S}_2\|_C + \|y - \bar{S}_2\|_C$$
$$\leqslant 2\tau_2(y''; 2h)_L + K_4 h^2$$

where

$$K_4 \leqslant \frac{(\|q\|_c + \frac{49}{4})}{2}(\|q\|_c \|y\|_c + \|f\|_c) \tag{7.66}$$

(since $\|y''\|_c \leqslant \|q\|_c \|y\|_c + \|f\|_c$).

We have obtained the following theorem.

Theorem 7.4. *Using the method of quadratic collocation splines* (7.45), (7.46) *for the numerical solution of problem* (7.44) *with assumption* (7.52), *the following estimate holds*:

$$\|y - S_2\|_{C[0,1]} \leqslant 2\tau_2(y''; 2h)_L + K_4 h^2$$

where the constant K_4 can be estimated by (7.66).

From this theorem we can obtain some corollaries, using the properties of $\tau_2(y''; \delta)_L$.

Corollary 7.6. *If y'' is Riemann integrable, then*

$$\lim_{n \to \infty} \|y - S_2\|_C = 0$$

Corollary 7.7. *If y'' has bounded variation, then*

$$\|y - S_2\|_C = O(h)$$

and if y''' has bounded variation, then

$$\|y - S_2\|_C = O(h^2)$$

Remark. Corollary 7.7 improves the statement that $\|y - S_2\|_C = O(h^2)$ if $y^{(iv)}$ is a bounded function in $[0, 1]$.

The estimate from Theorem 7.4 improves the estimate of Stečkin and Subbotin [1] (p. 221):

$$\|y - S_2\|_C \leqslant \tfrac{3}{16}\omega(y''; h)_C + O(h^2)$$

since in the general case $\tau_2(g; \delta)_L$ has better order than $\omega(g; \delta)_C$.

Let us now consider in a similar way the numerical solution of equation (7.44) by the method of cubic spline collocation.

Let S_3 be a cubic spline with knots at the points $x_i = i/n$, $i = -2, -1, \ldots,$ $n + 2$. Then S_3 has the representation (see Section 4.3.3.)

$$S_3(x) = \sum_{k=-1}^{n+1} d_k B_3(x - kh) \tag{7.67}$$

where

$$B_3(x) = \frac{1}{6h^3} \sum_{j=0}^{4} (-1)^j \binom{4}{j}(x + 2h - jh)_+^3 \tag{7.68}$$

We determinate the coefficients d_k, $k = -1, 0, \ldots, n + 1$, in representation (7.67) from the collocation and the boundary conditions:

$$S_3''(x_i) - q_i S_3(x_i) = f_i \quad i = 0, 1, \ldots, n$$
$$S_3(0) = \alpha, \quad S_3(1) = \beta \tag{7.69}$$

In view of the properties of the cubic B-spline B_3 (see Section 4.59), representation (7.67) and condition (7.69), we obtain the following system of linear algebraic equations for the coefficients d_i, $i = 0, 1, \ldots, n$ (after excluding d_{-1} and d_{n+1} as in the case of quadratic splines):

$$d_0 = y_0 - \frac{h^2(q_0 y_0 + f_0)}{6}$$

$$\Delta_h^2 d_{i-1} - \frac{6h^2 q_i}{6 - h^2 q_i} d_i = \frac{6h^2}{6 - h^2 q_i} f_i \quad i = 1, 2, \ldots, n-1 \tag{7.70}$$

$$d_n = y_n - \frac{h^2(q_n y_n + f_n)}{6}$$

We shall proceed as in the case of quadratic splines. Let

$$\bar\theta_i = -h^2[v_1 + ih(v_2 - v_1)] \quad i = 0, 1, \ldots, n \tag{7.71}$$

where

$$v_1 = \frac{q_0 y_0 + f_0}{6}, \quad v_2 = \frac{q_n y_n + f_n}{6} \tag{7.72}$$

For $w_i = d_i - \theta_i$, $i = 0, 1, \ldots, n$, we obtain the following system of linear algebraic equations:

$$w_0 = y_0$$

$$\Delta_h^2 w_{i-1} - \frac{6q_i}{6 - h^2 q_i} w_i = \frac{6}{6 - h^2 q_i} f_i + \frac{6q_i}{6 - h^2 q_i} \bar\theta_i \tag{7.73}$$

$$w_n = y_n$$

For $\bar\theta_i$ we have the estimate (see (7.71), (7.72))

$$|\bar\theta_i| = |d_i - w_i|$$
$$\leqslant h^2 \max\{|v_1|, |v_2|\}$$
$$\leqslant \frac{h^2(\|q\|_c \|y\|_c + \|f\|_c)}{6} \tag{7.74}$$

We shall now assume again, that h is sufficiently small, more precisely that

$$6 - h^2 \|q\|_c > 1, \quad \frac{6}{6 - h^2 \|q\|_c} \leqslant 2 \tag{7.75}$$

Let

$$\frac{6}{6 - h^2 q_i} = 1 + \beta_i \leqslant 2 \quad i = 1, 2, \ldots, n \tag{7.76}$$

Then

$$0 \leqslant \beta_i < h^2 \| q \|_C$$

Using this notation, the system (7.73) becomes

$$w_0 = y_0$$

$$\frac{\Delta_h^2 w_{i-1}}{h^2} + (1 + \beta_i) q_i w_i = (1 + \beta_i) f_i + (1 + \beta_i) q_i \bar{\theta}_i \quad i = 1, 2, \dots, n-1 \quad (7.77)$$

$$w_n = y_n$$

Since the exact solution of (7.44) satisfies equations (7.51), from (7.51) and (7.77) we obtain the equation for $Q_i = y_i - w_i$:

$$(Q_{\bar{x}})_x - pQ = \psi + F$$

$$Q_0 = Q_n = 0$$

where

$$p_i = (1 + \beta_i) q_i \geqslant 0, \quad F_i = -\beta_i y_i'' - (1 + \beta_i) q_i \bar{\theta}_i, \quad \psi_i = \frac{\Delta_h^2 y_{i-1}}{h^2} - y_i''$$

For this system we again apply Lemma 7.1. Using Lemma 7.8 and equations (7.74) and (7.76), we obtain

$$\| Q \|_{C,h} = \max \{ |y_i - w_i| : 0 \leqslant i \leqslant n \}$$

$$\leqslant \sum_{i=1}^{n-1} h^2 [|\beta_i| |y_i''| + (1 + \beta_i) |q_i| |\bar{\theta}_i|] + \sum_{i=1}^{n-1} h |\psi_i|$$

$$\leqslant h^2 \| q \|_C \| y'' \|_C + 2 \| q \|_C \frac{h^2}{6} (\| q \|_C \| y \|_C + \| f \|_C) + \frac{1}{2} \sum_{i=1}^{n-1} h w_2(y'', x_i; h)$$

$$\leqslant h^2 [\tfrac{4}{3} \| q \|_C (\| q \|_C \| y \|_C + \| f \|_C)] + \sum_{i=1}^{n-1} \frac{1}{2} \int_{x_{i-1/2}}^{x_{i+1/2}} w_2(y'', x; 2h) \, dx$$

$$= \tfrac{1}{2} \tau_2(y'', 2h)_L + K h^2 \tag{7.78}$$

where the constant K satisfies the inequality

$$K \leqslant \tfrac{4}{3} \| q \|_C (\| q \|_C \| y \|_C + \| f \|_C)$$

From (7.74) and (7.78) we obtain

$$|y_i - d_i| \leqslant |y_i - w_i| + |w_i - d_i|$$

$$\leqslant \tfrac{1}{2} \tau_2(y''; 2h)_L + K h^2 + \frac{h^2}{6} (\| q \|_C \| y \|_C + \| f \|_C)$$

$$= \tfrac{1}{2} \tau_2(y''; 2h)_L + K_1 h^2 \tag{7.79}$$

where the constant K_1 satisfies the inequality

$$K_1 \leqslant \frac{8\|q\|_c + 1}{6}(\|q\|_c\|y\|_c + \|f\|_c)$$

Using the fact that

$$S_3(x_i) = \frac{d_{i-1} + 4d_i + d_{i+1}}{6}$$

we obtain from (7.79)

$$|y_i - S_3(x_i)| = \left| y_i - \frac{d_{i-1} + 4d_i + d_{i+1}}{6} \right|$$

$$\leqslant \tfrac{1}{6}(|y_{i-1} - d_{i-1}| + 4|y_i - d_i| + |y_{i+1} - d_{i+1}|) + \frac{|\Delta_h^2 y_{i-1}|}{6}$$

$$\leqslant \tfrac{1}{2}\tau_2(y''; 2h)_L + K_1 h^2 + \tfrac{1}{6}w_2(y; h)_c \qquad (7.80)$$

Since

$$\omega_2(y; h)_c \leqslant h^2 \|y''\|_c \leqslant h^2(\|q\|_c\|y\|_c + \|f\|_c) \qquad (7.81)$$

then (7.80) and (7.81) finally give the following result.

Theorem 7.5. *Using the method of cubic spline collocation (7.69) for the numerical solution of equation (7.44) under condition (7.75), the following estimate holds:*

$$\|y - S_3\|_{c,h} = \max\{|y_i - S_3(x_i)| : 0 \leqslant i \leqslant n\} \leqslant \tfrac{1}{2}\tau_2(y''; 2h)_L + K_2 h^2$$

where the constant K_2 can be estimated by

$$K_2 \leqslant \tfrac{1}{3}(4\|q\|_c + 1)(\|q\|_c\|y\|_c + \|f\|_c) \qquad (7.82)$$

From Theorem 7.5 we obtain the following, as in the case of quadratic spline collocation, using Corollary 4.10 and Theorem 4.14.

Theorem 7.6. *Using the method of cubic spline collocation (7.69) for the numerical solution of equation (7.44) under condition (7.75) the following estimate holds:*

$$\|y - S_3\|_{C[0,1]} \leqslant c[\tau_2(y''; 2h)_L + K_2 h^2]$$

where c is an absolute constant, and K_2 is estimated by (7.82).

Corollary 7.8. *If y'' has bounded variation in the interval $[0, 1]$, then*

$$\|y - S_3\|_c = O(h)$$

and if y''' is a function of bounded variation in $[0, 1]$, then

$$\|y - S_3\|_c = O(h^2)$$

Theorem 7.6, is addition to Corollary 7.8, improves the known estimate of the error in the numerical solution of the boundary value problem by the method of collocation cubic splines (compare with Stećkin and Subbotin [1]).

7.4. The finite element method

Let us solve the equation

$$\frac{d}{dx}\left(k\frac{dy}{dx}\right) - qy = f \quad x \in [0, 1], \ y(0) = y(1) = 0 \tag{7.83}$$

by the method of finite elements (see, for example, Sendov and Popov [1]).

As usual, we shall assume that $k(x) \geqslant c_0 > 0$, $q(x) \geqslant 0$ for $x \in [0, 1]$.

Let us devide the interval $[0, 1]$ into n equal parts by means of the points $x_i = ih$, $i = 0, 1, \ldots, n$, $h = i/n$. We introduce the 'finite elements' φ_i, $i = 1, 2, \ldots$, $n - 1$:

$$\varphi_i(x) = \begin{cases} 0 & x < (i-1)/n \\ (x - x_{i-1})/h & x \in [x_{i-1}, x_i] \\ (x_{i+1} - x)/h & x \in [x_i, x_{i+1}] \\ 0 & x > (i+1)/n \end{cases} \tag{7.84}$$

It is easy to see, that the functions φ_i satisfy the equations

$$\int_0^1 \varphi_i(x)\varphi_j(x)\,dx = \begin{cases} 0 & j \leqslant i - 2, \\ h/6 & j = i - 1, \\ 2h/3 & j = i, \\ h/6 & j = i + 1, \\ 0 & j \geqslant i + 2. \end{cases} \tag{7.85}$$

We shall find the approximate solution $\tilde{y}$ of problem (7.83) in the form

$$\tilde{y}(x) = \sum_{i=1}^{n-1} a_i \varphi_i(x)$$

It follows immediately from the properties of the functions φ_i in (7.84) that $a_i = \tilde{y}(x_i) = \tilde{y}_i$.

The coefficients a_i (or $\tilde{y}_i$) will be obtained from the conditions

$$\int_0^1 \left\{ k\frac{d\tilde{y}}{dx}\frac{d\varphi_i}{dx} + (q\tilde{y} + f)\varphi_i \right\} dx = 0 \quad i = 1, 2, \ldots, n - 1 \tag{7.86}$$

From the definition of the functions φ_i it follows that

$$\frac{d\varphi_i}{dx} = \begin{cases} 0 & x \bar{\in} [x_{i-1}, x_{i+1}] \\ 1/h & x \in (x_{i-1}, x_i) \\ -1/h & x \in (x_i, x_{i+1}) \end{cases}$$

Since

$$\frac{d\tilde{y}}{dx} = \sum_{j=1}^{n-1} \tilde{y}_j \frac{d\varphi_j}{dx}$$

then for $i = 1, 2, \ldots, n$ we obtain

$$\int_0^1 k \frac{d\tilde{y}}{dx} \frac{d\varphi_i}{dx} dx = \frac{1}{h} \int_{x_{i-1}}^{x_i} k \left(\sum_{j=1}^{n-1} \tilde{y}_j \frac{d\varphi_j}{d\tilde{x}} \right) dx - \frac{1}{h} \int_{x_i}^{x_{i+1}} k \left(\sum_{j=1}^{n-1} \tilde{y}_j \frac{d\varphi_j}{dx} \right) dx$$

$$= \int_{x_{i-1}}^{x_i} k(x) \frac{y_i - y_{i-1}}{h^2} dx - \int_{x_i}^{x_{i+1}} k(x) \frac{y_{i+1} - y_i}{h^2} dx$$

$$= \left[\tilde{k}_i \left(\frac{y_i - y_{i-1}}{h} \right) - \tilde{k}_{i+1} \left(\frac{y_{i+1} - y_i}{h} \right) \right] \bigg/ h$$

$$= -(\tilde{k} y_{\tilde{x}})_{x,i} \tag{7.87}$$

where we have set

$$\tilde{k}_i = \int_{x_{i-1}}^{x_i} k(x) \, dx$$

In an analogous way we have

$$\int_0^1 q\tilde{y}\varphi_i \, dx = \int_0^1 q \sum_{j=1}^{n-1} \tilde{y}_j \varphi_j \varphi_i \, dx$$

$$= \tilde{y}_{i-1} \int_{x_{i-1}}^{x_i} q\varphi_{i-1}\varphi_i \, dx + \tilde{y}_i \left(\int_{x_{i-1}}^{x_{i+1}} q\varphi_i\varphi_i \, dx \right) + \tilde{y}_{i+1} \int_{x_i}^{x_{i+1}} q\varphi_i\varphi_{i+1} \, dx$$

$$= \tilde{q}_{i,i-1}\tilde{y}_{i-1} + \tilde{q}_{i,i}\tilde{y}_i + \tilde{q}_{i,i+1}\tilde{y}_{i+1} \tag{7.88}$$

where we use the notation

$$\tilde{q}_{i,i-1} = \int_{x_{i-1}}^{x_i} q\varphi_{i-1}\varphi_i \, dx$$

$$\tilde{q}_{i,i} = \int_{x_{i-1}}^{x_{i+1}} q\varphi_1^2 \, dx$$

$$\tilde{q}_{i,i+1} = \int_{x_i}^{x_{i+1}} q\varphi_i\varphi_{i+1} \, dx \quad (\varphi_0 = \varphi_n = 0)$$

Using (7.87) and (7.88), the system (7.86) can be rewritten in the form

$$(\tilde{k}\tilde{y}_{\tilde{x}})_{x,i} - (\tilde{y}_{i-1}\tilde{q}_{i,i-1} + \tilde{y}_i\tilde{q}_{i,i} + \tilde{y}_{i+1}\tilde{q}_{i,i+1}) = \tilde{f}_i \quad i = 1, 2, \ldots, n-1 \tag{7.89}$$

where

$$\tilde{f}_i = \int_0^1 f\varphi_i \, dx$$

and $\tilde{k}$ and $\tilde{q}$ are given by (7.87) and (7.88).

The system (7.89) together with the boundary conditions $\tilde{y}_0 = \tilde{y}_n = 0$ give us a system of $n-1$ linear algebraic equations for the unknowns $\tilde{y}_i$, $i = 1, 2, \ldots, n-1$.

To estimate the error $\varepsilon = y - \tilde{y}$ between the exact solution y of (7.83) and the solution $\tilde{y}$ of (7.89), let us derive a system of linear algebraic equations similar to (7.89) which is satisfied by the values of the exact solution y_i, $i = 0, 1, \ldots, n$. If we multiply (7.83) by φ_i and integrate from 0 to 1, we obtain

$$\int_0^1 \frac{d}{dx}\left(k\frac{dy}{dx}\right)\varphi_i\,dx - \int_0^1 qy\varphi_i\,dx = \int_0^1 f\varphi_i\,dx \quad i = 1, 2, \ldots, n-1 \quad (7.90)$$

For the first integral, integrating by parts and in view of the equations for $d\varphi_i/dx$ and (7.87). We obtain

$$\int_0^1 \varphi_i\,d\left(k\frac{dy}{dx}\right) = -\int_0^1 k\frac{dy}{dx}\frac{d\varphi_i}{dx}\,dx$$

$$= -\frac{1}{h}\int_{x_{i-1}}^{x_i} k\frac{dy}{dx}\,dx + \frac{1}{h}\int_{x_i}^{x_{i+1}} k\frac{dy}{dx}\,dx$$

$$= (\tilde{k}y_{\tilde{x}})_{x,i} - \frac{1}{h}\left[\int_{x_{i-1}}^{x_i}\left(k - \frac{1}{h}\tilde{k}_i\right)\frac{dy}{dx}\,dx - \int_{x_i}^{x_{i+1}}\left(k - \frac{1}{h}\tilde{k}_{i+1}\right)\frac{dy}{dx}\,dx\right]$$

$$= (\tilde{k}y_{\tilde{x}})_{x,i} + r_i$$

where

$$r_i = -\frac{1}{h}\left[\int_{x_{i-1}}^{x_i}\left(k - \frac{1}{h}\tilde{k}_i\right)\frac{dy}{dx}\,dx - \int_{x_i}^{x_{i+1}}\left(k - \frac{1}{h}\tilde{k}_{i+1}\right)\frac{dy}{dx}\,dx\right]$$

We shall estimate r_i in two ways.

(a) We have

$$r_i = -\frac{1}{h}\left\{\int_{x_{i-1}}^{x_i}(k - k_{i-1/2})\frac{dy}{dx}\,dx + \int_{x_{i-1}}^{x_i}\left(k_{i-1/2} - \frac{1}{h}\tilde{k}_{i+1}\right)\frac{dy}{dx}\,dx\right.$$
$$\left. -\left[\int_{x_i}^{x_{i+1}}(k - k_{i+1/2})\frac{dy}{dx}\,dx + \int_{x_i}^{x_{i+1}}\left(k_{i+1/2} - \frac{1}{h}\tilde{k}_{i+1/2}\right)\frac{dy}{dx}\,dx\right]\right\} \quad (7.91)$$

Since

$$\left|k_{i-1/2} - \frac{1}{h}\tilde{k}_i\right| = \left|\frac{1}{h}\int_{x_{i-1}}^{x_i}[k_{i-1/2} - k(x)]\,dx\right| \leqslant \omega_2\left(k, x_{i-1/2}; \frac{h}{2}\right)$$

$$\left|k_{i+1/2} - \frac{1}{h}\tilde{k}_{i+1}\right| = \left|\frac{1}{h}\int_{x_i}^{x_{i+1}}[k_{i+1/2} - k(x)]\,dx\right| \leqslant \omega_2\left(k, x_{i+1/2}; \frac{h}{2}\right)$$

applying Lemma 7.9 for the first and the third integrals in (7.91), we obtain

$$|r_i| \leqslant \|y'\|_C \omega_2\left(k, x_{i-1/2}; \frac{h}{2}\right)$$

$$+ \omega\left(k, x_{i-1/2}; \frac{h}{2}\right)\omega\left(y', x_{i-1/2}; \frac{h}{2}\right)$$

$$+ \|y'\|_C \omega_2 \left(k, x_{i-1/2}; \frac{h}{2} \right) + \|y'\|_C \omega_2 \left(k, x_{i+1/2}; \frac{h}{2} \right)$$

$$+ \omega \left(k, x_{i+1/2}; \frac{h}{2} \right) \omega \left(y', x_{i+1/2}; \frac{h}{2} \right)$$

$$+ \|y'\|_C \omega_2 \left(k_1 x_{i+1/2}; \frac{h}{2} \right)$$

i.e.

$$|r_i| \leqslant 2\|y'\|_C \left[\omega_2 \left(k, x_{i-1/2}; \frac{h}{2} \right) + \omega_2 \left(k, x_{i+1/2}; \frac{h}{2} \right) \right]$$

$$+ \omega \left(k, x_{i-1/2}; \frac{h}{2} \right) \omega \left(y', x_{i-1/2}; \frac{h}{2} \right) + \omega \left(k, x_{i+1/2}; \frac{h}{2} \right)$$

$$\times \omega \left(y', x_{i+1/2}; \frac{h}{2} \right) \tag{7.92}$$

(b) In the case when the function k is not smooth enough, we can obtain another estimate. Let

$$(\tilde{k}y')_i = \int_{x_{i-1}}^{x_i} ky' \, dx$$

Then

$$\int_{x_{i-1}}^{x_i} \left(k - \frac{1}{h}\tilde{k}_i \right) y' \, dx = \int_{x_{i-1}}^{x_i} \left[ky' - \frac{1}{h}(\tilde{k}y')_i \right] dx + \int_{x_{i-1}}^{x_i} \frac{1}{h} [(\tilde{k}y')_i - \tilde{k}_i y'] \, dx$$

$$= \frac{1}{h} \int_{x_{i-1}}^{x_i} \int_{x_{i-1}}^{x_i} [(ky')(x) - (ky')(u)] \, du \, dx$$

$$+ \frac{1}{h} \int_{x_{i-1}}^{x_i} \int_{x_{i-1}}^{x_i} [(k(u)y'(u) - k(u)y'(x)] \, du \, dx \tag{7.93}$$

By change of variables we see that for every function g we have

$$\left| \frac{1}{h} \int_{x_{i-1}}^{x_i} \int_{x_{i-1}}^{x_i} [g(u) - g(x)] \, du \, dx \right| \leqslant \frac{1}{h} \int_{-h}^{h} \int_{x_{i-1}}^{x_i} |g(u + x) - g(x)| \, dx \, du$$

(if the argument of the integrand in the second integral is out side the interval $[0, 1]$, we assume that these values are zero).

Therefore from (7.93) we obtain

$$\left| \int_{x_{i-1}}^{x_i} \left(k - \frac{1}{h}\tilde{k}_i \right) y'_i \, dx \right| \leqslant \frac{1}{h} \int_{-h}^{h} \int_{x_{i-1}}^{x_i} |(ky')(x + u) - (ky')(x)| \, dx \, du$$

$$+ \|k\|_C \frac{1}{h} \int_{-h}^{h} \int_{x_{i-1}}^{x_i} |y'(x + u) - y'(x)| \, dx \, du$$

In an analogous way we see that

$$\left| \int_{x_i}^{x_{i+1}} \left(k - \frac{1}{h}\tilde{k}_{i+1} \right) y' \, dx \right| \leq \frac{1}{h} \int_{-h}^{h} \int_{x_i}^{x_{i+1}} |(ky')(x+u) - (ky')(x)| \, dx \, du$$

$$+ \|k\|_c \frac{1}{h} \int_{-h}^{h} \int_{x_i}^{x_{i+1}} |y'(x+u) - y'(x)| \, dx \, du$$

and therefore

$$|r_i| \leq \frac{1}{h^2} \int_{-h}^{h} \int_{x_{i-1}}^{x_{i+1}} |(ky')(x+u) - (ky')(x)| \, dx \, du$$

$$+ \frac{\|k\|_c}{h^2} \int_{-h}^{h} \int_{x_{i-1}}^{x_{i+1}} |y'(x+u) - y'(x)| \, dx \, du \qquad (7.94)$$

Let us now set

$$\theta_i = \int_0^1 qy\varphi_i \, dx - (\tilde{q}_{i,i-1}y_{i-1} + \tilde{q}_{i,i}y_i + \tilde{q}_{i,i+1}y_{i+1}) \qquad (7.95)$$

Then system (7.90), in view of r_i and (7.95), takes the form

$$(\tilde{k}y_{\bar{x}})_{x,i} - (\tilde{q}_{i,i-1}y_{i-1} + \tilde{q}_{i,i}y_i + \tilde{q}_{i,i+1}y_{i+1})$$

$$= \int_0^1 f\varphi_i \, dx - r_i + \theta_i \quad i = 1, 2, \ldots, n-1, \ y_0 = y_n = 0 \qquad (7.96)$$

From (7.89) and (7.96) we see that the error $\varepsilon = y - \tilde{y}$ satisfies the following system of linear algebraic equations:

$$(\tilde{k}\varepsilon_{\bar{x}})_{x,i} - (\tilde{q}_{i,i-1}\varepsilon_{i-1} + \tilde{q}_{i,i}\varepsilon_i + \tilde{q}_{i,i+1}\varepsilon_{i+1}) = \theta_i + r_i \qquad (7.97)$$

where $i = 1, 2, \ldots, n-1$, $\varepsilon_0 = \varepsilon_n = 0$.

From Lemma 7.6 it follows that if h is sufficiently small so that

$$\frac{c_0}{h^2} > \frac{2\|q\|_c}{h} \geq \max_{i,j} |\tilde{q}_{i,j}| \qquad (7.98)$$

then

$$\|\varepsilon\|_{c,k} \leq \frac{4}{c_0} \sum_{i=1}^{n-1} \left(h|r_i| + \left| \sum_{j=1}^{i-1} h\theta_j \right| \right) \qquad (7.99)$$

Let us estimate

$$\sum_{j=1}^{i-1} \theta_j$$

using the definition of θ_j from (7.95).

Let

$$\bar{y} = \sum_{j=1}^{n-1} y_j\varphi_j$$

Then $\bar{y}$ is the interpolation spline of first degree for the solution y, constructed using the points x_i, $i = 0, 1, \ldots, n$. Therefore, by Theorem 4.4, we have

$$\|y - \bar{y}\|_L \leqslant 2\tau_2(y; h)_L \tag{7.100}$$

Since

$$\left|\sum_{j=1}^{i-1} \varphi_j\right| \leqslant 1$$

then from (7.100), since

$$\int_0^1 q\bar{y}\varphi_i \, dx = \tilde{q}_{i,i-1} y_{i-1} + \tilde{q}_{i,i} y_i + \tilde{q}_{i,i+1} y_{i+1}$$

we obtain

$$\left|\sum_{j=1}^{i-1} h\theta_j\right| = h\left|\sum_{j=1}^{i-1} \theta_j\right|$$

$$= h\left|\sum_{j=1}^{i-1}\left(\int_0^1 qy\varphi_j \, dx - \int_0^1 q\bar{y}\varphi_j \, dx\right)\right|$$

$$\leqslant h\|q\|_C \|y - \bar{y}\|_{L[0,1]}$$

$$\leqslant 2h\|q\|_C \tau_2(y; h)_L$$

Therefore the estimate (7.99) becomes

$$\|\varepsilon\|_{C,h} \leqslant \frac{4}{C_0}\left[2\|q\|_C \tau_2(y; h)_L + \sum_{i=1}^{n-1} h|r_i|\right] \tag{7.101}$$

If we take into account all the different estimates for $|r_i|$, we obtain for case (a) from (7.92)

$$\sum_{i=1}^{n-1} h|r_i| \leqslant 2\|y'\|_C\left\{\sum_{i=1}^{n-1} h\left[\omega_2\left(k, x_{i-1/2}; \frac{h}{2}\right) + \omega_2\left(k, x_{i+1/2}; \frac{h}{2}\right)\right]\right\}$$

$$+ \sum_{i=1}^{n-1}\left[\omega\left(k, x_{i-1/2}; \frac{h}{2}\right)\omega\left(y', x_{i-1/2}; \frac{h}{2}\right)h\right.$$

$$\left. + \omega_2\left(k, x_{i+1/2}; \frac{h}{2}\right)\omega\left(y', x_{i+1/2}; \frac{h}{2}\right)h\right]$$

$$\leqslant 2\|y'\|_C\left\{\sum_{i=1}^{n-1}\left[\int_{x_{i-1}}^{x_i} \omega_2(k, x; h) \, dx + \int_{x_i}^{x_{i+1}} \omega_2(k, x; h) \, dx\right]\right\}$$

$$+ \sum_{i=1}^{n-1} h\left[\omega\left(k, x_{i-1/2}; \frac{h}{2}\right)\omega\left(y', x_{i-1/2}; \frac{h}{2}\right)\right.$$

$$\left. + \omega\left(k, x_{i+1/2}; \frac{h}{2}\right)\omega\left(y', x_{i+1/2}; \frac{h}{2}\right)\right] \tag{7.102}$$

We have

$$\sum_{i=1}^{n-1} h\left[\omega\left(k, x_{i-1/2}; \frac{h}{2}\right)\omega\left(y', x_{i-1/2}; \frac{h}{2}\right) + \omega\left(k_2 \, x_{i+1/2}; \frac{h}{2}\right)\omega\left(y', x_{i+1/2}; \frac{h}{2}\right)\right]$$

$$\leqslant \omega\left(k; \frac{h}{2}\right)_C \left\{ \sum_{i=1}^{n-1} \left[\int_{x_{i-1}}^{x_i} \omega(y', x; h)\, dx + \int_{x_i}^{x_{i+1}} \omega(y', x; h)\, dx \right] \right\} \tag{7.103}$$

and

$$\sum_{i=1}^{n-1} h\left[\omega\left(k, x_{i-1/2}; \frac{h}{2}\right)\omega\left(y', x_{i-1/2}; \frac{h}{2}\right) + \omega\left(k, x_{i+1/2}; \frac{h}{2}\right)\omega\left(y', x_{i+1/2}; \frac{h}{2}\right)\right]$$

$$\leqslant \omega\left(y'; \frac{h}{2}\right)_C \left\{ \sum_{i=1}^{n-1} \left[\int_{x_{i-1}}^{x_i} \omega(k, x; h)\, dx + \int_{x_i}^{x_{i+1}} \omega(k, x; h)\, dx \right] \right\} \tag{7.104}$$

From (7.102), (7.103) and (7.104) we obtain the following two estimate for

$$\sum_{i=1}^{n-1} h |r_i|$$

in case (a):

$$\sum_{i=1}^{n-1} h |r_i| = 4 \| y' \|_C \tau_2(k; h)_L + 2\omega\left(k; \frac{h}{2}\right)_C \tau(y'; h)_L \tag{7.105}$$

$$\sum_{i=1}^{n-1} h |r_i| \leqslant 4 \| y' \|_C \tau_2(k; h)_L + 2\omega\left(y'; \frac{h}{2}\right)_C \tau(k; h)_L \tag{7.106}$$

From (7.94) we have

$$\sum_{i=1}^{n-1} h |r_i| \leqslant \frac{1}{h}\int_{-h}^{h} \sum_{i=1}^{n-1}\left[\int_{x_{i-1}}^{x_i} |(ky')(x+u) - (ky')(x)|\, dx\, du \right]$$

$$+ \frac{\| k \|_C}{h}\int_{-h}^{h}\sum_{i=1}^{n-1}\int_{x_{i-1}}^{x_i} |y'(x+u) - y'(x)|\, dx\, du$$

$$\leqslant 4\omega(ky'; h)_L + 4 \| k \|_C \omega(y'; h)_L \tag{7.107}$$

The estimates (7.101), (7.105)–(7.107) gives us the following.

Theorem 7.7. *For the finite element method (7.89) for the solution of problem (7.83) under condition (7.98), the following estimates for $\varepsilon = y - \tilde{y}$ hold:*

$$\| \varepsilon \|_{C,h} = \max \{ |y_i - \tilde{y}_i| : 0 \leqslant i \leqslant n \}$$

$$\leqslant \frac{8}{c_0}\left[\| q \|_C \tau_2(y; h)_L + 2 \| y' \|_C \tau_2(k; h)_L + \omega\left(k; \frac{h}{2}\right)_C \tau(y'; h)_L \right]$$

$$\| \varepsilon \|_{C,h} \leqslant \frac{8}{c_0}\left[\| q \|_C \tau_2(y; h)_L + 2 \| y' \|_C \tau_2(k; h)_L + \omega\left(y'; \frac{h}{2}\right)_C \tau(k; h)_L \right]$$

$$\| \varepsilon \|_{C,h} \leqslant \frac{8}{c_0}\{ \| q \|_C \tau_2(y; h)_L + 2[\omega(ky'; h)_L + \| k \|_C \omega(y'; h)_L] \}$$

Again from Theorem 7.7 we can derive many consequences. We shall note first the case when k is constant. Then we obtain the following.

Corollary 7.9 *If* $k = c_0$ *(constant), then*

$$\|\varepsilon\|_{C,h} \leqslant \frac{8}{c_0} \|q\|_C \tau_2(y; h)_L$$

If q and f are bounded measurable functions, then y'' is also a bounded measurable functions and therefore $\tau_2(y; h)_L = O(h^2)$, i.e. in this case $\|\varepsilon\|_{C,h} = O(h^2)$. This result is obtained in another way by Fedotova [1].

Notice that for the difference $\varepsilon = y - \tilde{y}$ we can obtain, using the estimate for $\|\varepsilon\|_{C,h}$, an estimate for every $x \in [0, 1]$. In fact, $\tilde{y}$ is a piecewise linear function with knots in the points x_i, $i = 0, 1, \ldots, n$. Let

$$\bar{y} = \sum_{i=1}^{n-1} y_i \varphi_i$$

be the interpolating spline for y of first degree. By Theorem 4.4 we have

$$\|y - \bar{y}\|_{C[0,1]} \leqslant 2\omega_2(y; h)_C$$

Therefore

$$\|y - \tilde{y}\|_{C[0,1]} \leqslant \|y - \bar{y}\|_{C[0,1]} + \|\bar{y} - \tilde{y}\|_{C[0,1]}$$
$$\leqslant 2\omega_2(y; h)_L + \|y - \tilde{y}\|_{C,h}$$

from which we obtain the following theorem.

Theorem 7.8. *For the finite element method* (7.89) *for the solution of problem* (7.83) *under the assumption* (7.98) *the following estimate holds*:

$$\|y - \tilde{y}\|_{C[0,1]} = \max\{|y(x) - \tilde{y}(x)| : x \in [0, 1]\} \leqslant \|\varepsilon\|_{C,h} + 2\omega_2(y; h)_C$$

where $\|\varepsilon\|_{C,h}$ *is estimated by Theorem 7.7.*

Again, if q and f are bounded measurable functions and k is constant, we have $\omega_2(y; h)_C = O(h^2)$ and therefore in this case we again have $\|y - \tilde{y}\|_C = O(h^2)$.

In practice the integrals for $\tilde{q}$ and $\tilde{f}$ are evaluated by means of some quadrature formulae, giving additional errors. Since the errors of the quadrature formulae can be estimated by the averaged moduli or by the local moduli, in each concrete case we can obtain estimates, by means of the averaged moduli of smoothness.

7.5. Notes

The results from Sections 7.2 and 7.3 are obtained by Andreev *et al.* [3]. The results from Section 7.4 are not published, except for Theorem 7.7 (see Popov [2]).

Alexandrov [1] applies multivariate averaged moduli of smoothness to estimate the error of the numerical solution of elliptic equations.

8 ONE-SIDED APPROXIMATION OF FUNCTIONS

In this chapter we shall consider the problem of one-sided approximation of functions. The theory of one-sided approximation of functions by means of trigonometric polynomials, spline functions and algebraic polynomials has undergone intensive development in the last few years (see Babenko and Ligun [1], Doronin and Ligun [1], Popov [1], Andreev *et al.* [1], Freud [1], Ganelius [1], Freud and Popov [1] and the references therein). One of the new facts in this theory is that analogues of the classical theorems of Jackson, Bernstein, Salem–Stečkin for one-sided approximations are obtained, not by means of the classical moduli of continuity and smoothness, but by using the averaged moduli of smoothness (see Andreev *et al.* [1] [2], Popov and Andreev [1], Popov [3]).

We shall prove here direct and converse theorems for best one-sided approximation by means of trigonometric polynomials. In Section 8.1 we give some classical results for best approximation by means of trigonometric polynomials. In Section 8.2 we prove direct theorems for best one-sided trigonometric approximation, which are analogous to the theorems of Jackson and Stečkin. In Section 8.3 we prove a converse theorem for best trigonometric one-sided approximation, which is analogous to the theorems of Bernstein and Salem–Stečkin. Both theorems together give us a characterization of the order of best one-sided trigonometric approximations by means of the averaged moduli of smoothness.

8.1. Classical theorems for best approximation of functions by means of trigonometric polynomials

In what follows we shall consider 2π-periodic functions. The set of all trigonometric polynomials of order n will be denoted by T_n, i.e. $t \in T_n$ if

$$t(x) = a_0 + \sum_{k=1}^{n} (a_k \cos kx + b_k \sin kx)$$

162

Let the 2π-periodic function f belong to $L_p[0, 2\pi]$. The best approximation of f by trigonometric polynomials of order n in $L_p[0, 2\pi]$, $1 \leqslant p \leqslant \infty$, is defined by

$$E_n(f)_p = E_n(f)_{L_p} = \inf\{\|f - t\|_p : t \in T_n\}$$

and this trigonometric polynomial, for which $E_n(f)_p$ is attained, is called the trigonometric polynomial of order n of the best L_p-approximation of f. It is well known that for $p \geqslant 1$ such a polynomial exists and for $p > 1$ this polynomial is unique. In the case $p = \infty$ we sometimes use the term uniform approximation.

The classical Jackson theorem is as follows:

$$E_n(f)_{C[0, 2\pi]} \leqslant c\omega(f; n^{-1})_{C[0, 2\pi]}$$

where c is an absolute constant (see Natanson [1], Dzjadik [1] Lorentz [1]).

At a later stage this theorem is generalized for L_p-metrics ($1 \leqslant p \leqslant \infty$) and for arbitrary moduli of smoothness by Steckin [1].

$$E_n(f)_{L_p} \leqslant c(k)\omega_k(f; n^{-1})_{L_p} \quad n > k \tag{8.1}$$

We see that in this direct theorem the best approximation is estimated by means of the corresponding modulus of smoothness. In the converse theorems of approximation theory the moduli of smoothness are estimated by means of best approximations. Without formulating the classical Bernstein theorem in its first form, we shall give here its modification, due to Salem–Steckin:

$$\omega_k(f; n^{-1})_{Lp} \leqslant \frac{c(k)}{n^k} \sum_{s=0}^{n} (s + 1)^{k-1} E_s(f)_{L_p} \tag{8.2}$$

From theorems given in (8.1) and (8.2) it is possible to obtain a characterization of the best trigonometric approximation by means of the corresponding moduli of smoothness in L_p. For example, for (8.1) and (8.2) it follows easily that

$$\omega_k(f; \delta)_{L_p} = O(\delta^\alpha) \Leftrightarrow E_n(f)_{L_p} = O(n^{-\alpha}) \quad 0 < \alpha < k \tag{8.3}$$

Our aim is to obtain analogues of (8.1)–(8.3) for best one-sided approximations.

8.2. Direct theorems for the best one-sided approximation

Let f be a bounded measurable 2π-periodic function. The best one-sided approximation of f by means of trigonometric polynomials of order n in L_p is given by

$$\tilde{E}_n(f)_{L_p} = \inf\{\|P - Q\|_{L_p} : P, Q \in T_n, Q(x) \leqslant f(x) \leqslant P(x) \text{ for every } x\} \tag{8.4}$$

The first non-trivial estimates for best one-sided approximation are due to Ganelius [1] (for the trigonometric case) and to Freud [1] (for the algebraic case) for $p = 1$. Notice that for $p = \infty$ the best one-sided approximation coincides up to a constant with the unrestricted best approximation, so far the one-sided approximation only the case $p < \infty$ is of interest.

First we shall obtain an estimate for $\tilde{E}_n(f)_{L_p}$ by $E_n(f')_{L_p}$, using a combination of the methods of Ganelius [1] and Popov [1].

Theorem 8.1. *Let f be an absolutely continuous function and $f' \in L_p[0, 2\pi]$. Then*

$$\tilde{E}_n(f)_{L_p} \leqslant \frac{2\pi}{n+1} E_n(f')_{L_p}$$

Proof. First note that by the assumptions of the theorem the function f has the representation

$$f(x) = \frac{1}{2\pi} \int_0^{2\pi} f(t)\, dt + \frac{1}{\pi} \int_0^{2\pi} B_1(x - t) f'(t)\, dt \qquad (8.5)$$

where B_1 is the first Bernoulli function:

$$B_1(x) = -\sum_{k=1}^{\infty} \frac{\sin kx}{k} = \begin{cases} \dfrac{x - \pi}{2} & 0 < x < 2\pi \\ \\ 0 & x = 0 \end{cases} \qquad (8.6)$$

Let us consider the trigonometric polynomials $T_{1,n} \in T_{n-1}$ and $t_{n,1} \in T_{n-1}$ determined by the following Hermite interpolation conditions:

$$T_{1,n}\left(\frac{2\pi}{n}k\right) = B_1\left(\frac{2\pi}{n}k\right) \quad k = 1, 2, \ldots, n-1,\; T_{1,n}(0) = \frac{\pi}{2}$$

$$T'_{1,n}\left(\frac{2\pi}{n}k\right) = B'_1\left(\frac{2\pi}{n}k\right) = \tfrac{1}{2} \quad k = 1, 2, \ldots, n-1 \qquad (8.7)$$

$$t_{1,n}\left(\frac{2\pi}{n}k\right) = B_1\left(\frac{2\pi}{n}k\right) \quad k = 1, 2, \ldots, n-1,\; t_{1,n}(0) = -\frac{\pi}{2}$$

$$t'_{1,n}\left(\frac{2\pi}{n}k\right) = B'_1\left(\frac{2\pi}{n}k\right) = \tfrac{1}{2} \quad k = 1, 2, \ldots, n-1 \qquad (8.8)$$

It is not difficult to verify that the graph of the polynomial $T_{1,n}$ is above the graph of B_1, and the graph of $t_{1,n}$ is under the graph of B_1, with touching points $2\pi k/n$, $k = 1, 2, \ldots, n-1$, and respectively 0 and 2π (if we accept that at these points the graph of B_1 is the interval $[-\pi/2, \pi/2]$). Therefore for $T_{1,n}$ and $t_{1,n}$ we have for every point x

$$t_{1,n}(x) \leqslant B_1(x) \leqslant T_{1,n}(x) \qquad (8.9)$$

The Hermite quadrature formula of Gaussian type (see Bahvalov [1]) states that if $T \in T_{n-1}$, then

$$\int_0^{2\pi} T(t)\, dt = \int_0^{2\pi} T(x + t)\, dt = \frac{2\pi}{n} \sum_{k=0}^{n-1} T\left(\frac{2\pi k}{n}\right)$$

Applying this formula for $Q = T_{1,n} - t_{1,n} \in T_{2n-1}$, from (8.7) and (8.8), we obtain

$$\int_0^{2\pi} Q(t)\,dt = \int_0^{2\pi} Q(x+t)\,dt$$

$$= \frac{2\pi}{n}\sum_{k=0}^{n-1}\left[T_{1,n}\left(\frac{2\pi}{n}k\right) - t_{1,n}\left(\frac{2\pi}{n}k\right)\right] = \frac{2\pi^2}{n}$$

From this and (8.9) we obtain

$$\frac{1}{\pi}\int_0^{2\pi}|T_{1,n}(t) - t_{1,n}(t)|\,dt = \frac{1}{\pi}\int_0^{2\pi}[T_{1,n}(t) - t_{1,n}(t)]\,dt = \frac{2\pi}{n} \qquad (8.10)$$

Let us return to representation (8.5). Let $q \in T_{n-1}$ be a trigonometric polynomial of best L_p approximation for f', i.e. $E_{n-1}(f')_{L_p} = \|f' - q\|_{L_p}$. For

$$P(x) = \frac{1}{\pi}\int_0^{2\pi} B_1(x-t)q(t)\,dt$$

it is easy to verify, using representation (8.6), that $P \in T_{n-1}$. Then from (8.5) we obtain

$$f(x) - P(x) = \frac{1}{2\pi}\int_0^{2\pi} f(t)\,dt + \frac{1}{\pi}\int_0^{2\pi} B_1(x-t)[f'(t) - q(t)]\,dt$$

Using the notation

$$S_+(t) = [f'(t) - q(t)]_+ = \begin{cases} f'(t) - q(t) & \text{if } f'(t) - q(t) \geqslant 0 \\ 0 & \text{if } f'(t) - q(t) < 0 \end{cases}$$

$$S_-(t) = f'(t) - q(t) - S_+(t)$$

then $S_+(t) \geqslant 0$, $S_-(t) \leqslant 0$, $S_+ + S_- = f' - q$.

Let us consider the trigonometric polynomials

$$T_+(x) = \frac{1}{\pi}\int_0^{2\pi} T_{1,n}(x-t)S_+(t)\,dt$$

$$T_-(x) = \frac{1}{\pi}\int_0^{2\pi} T_{1,n}(x-t)S_-(t)\,dt$$

$$t_+(x) = \frac{1}{\pi}\int_0^{2\pi} t_{1,n}(x-t)S_+(t)\,dt$$

$$t_-(x) = \frac{1}{\pi}\int_0^{2\pi} t_{1,n}(x-t)S_-(t)\,dt$$

Obviously $T_+, T_-, t_+, t_- \in T_{n-1}$ and

$$f(x) \leqslant R_1(x) = \frac{1}{2\pi}\int_0^{2\pi} f(t)\,dt + P(x) + T_+(x) + t_-(x)$$

$$f(x) \geqslant R_2(x) = \frac{1}{2\pi}\int_0^{2\pi} f(t)\,dt + P(x) + T_-(x) + t_+(x)$$

since

$$f(x) = \frac{1}{2\pi} \int_0^{2\pi} f(t)\,dt + P(x) + \frac{1}{\pi} \int_0^{2\pi} B_1(x-t)[S_+(t) + S_-(t)]\,dt$$

$$\leqslant \frac{1}{2\pi} \int_0^{2\pi} f(t)\,dt + P(x) + \frac{1}{\pi} \int_0^{2\pi} T_{1,n}(x-t)S_+(t)\,dt + \frac{1}{\pi} \int_0^{2\pi} t_1(x-t)S_-(t)\,dt$$

$$f(x) = \frac{1}{2\pi} \int_0^{2\pi} f(t)\,dt + P(x) + \frac{1}{\pi} \int_0^{2\pi} B_1(x-t)[S_+(t) + S_-(t)\,dt]\,dt$$

$$\geqslant \frac{1}{2\pi} \int_0^{2\pi} f(t)\,dt + P(x) + \frac{1}{\pi} \int_0^{2\pi} t_{1,n}(x-t)S_+(t)\,dt + \frac{1}{\pi} \int_0^{2\pi} T_{1,n}(x-t)S_-(t)\,dt$$

We have $R_1 \in T_{n-1}$, $R_2 \in T_{n-1}$. Let us now estimate $\|R_1 - R_2\|_{L_p}$. Using the generalized Minkovski inequality for $1 \leqslant p \leqslant \infty$

$$\left\| \int_0^{2\pi} |g(t)|\,|K(\cdot,t)|\,dt \right\|_{L_p} \leqslant \int_0^{2\pi} |g(t)|\,\|K(\cdot,t)\|_{L_p}\,dt$$

from (8.10) we obtain

$$\|R_1 - R_2\|_{L_p} = \|T_+ + t_- - T_- - t_+\|_{L_p}$$

$$= \left\| \frac{1}{\pi} \int_0^{2\pi} [T_{1,n}(x-t) - t_{1,n}(x-t)]|f'(t) - q(t)|\,dt \right\|_{L_p}$$

$$= \left\| \frac{1}{\pi} \int_0^{2\pi} |T_{1,n}(t) - t_{1,n}(t)|\,|f'(x+t) - q(x+t)|\,dt \right\|_{L_p}$$

$$\leqslant \frac{1}{\pi} \int_0^{2\pi} |T_{1,n}(t) - t_{1,n}(t)|\,\|f' - g\|_{L_p}\,dt = \frac{2\pi}{n} E_{n-1}(f')_{L_p} \quad \blacksquare$$

From Theorem 8.1 and (8.1) we obtain the following immediately.

Corollary 8.1. *Let $n > k$, where k is an integer. Then there exists a constant $c(k)$, dependent only on k, such that*

$$\tilde{E}_n(f)_{L_p} \leqslant c(k)n^{-1}\omega_k(f';n^{-1})_{L_p}$$

One other important corollary which we shall use is the following.

Corollary 8.2. *Let the function f have kth derivative $f^{(k)} \in L_p$. Then*

$$\tilde{E}_n(f)_{L_p} \leqslant c'(k)n^{-k}\|f^{(k)}\|_{L_p} \quad n > k$$

where the constant $c'(k)$ depends only on k.

Theorem 8.2. *Let f be a bounded measurable 2π-periodic function. Then*

$$\tilde{E}_n(f)_{L_p} \leqslant 120\tau(f;n^{-1})_{L_p} \quad 1 \leqslant p \leqslant \infty$$

Proof. Let $x_i = i\pi/n$, $i = 0, 1, \ldots, 2n$, $y_i = (x_{i-1} + x_i)/2$, $i = 1, 2, \ldots, 2n$,

$y_{2n+1} = y_1$, and let us define the 2π-periodic functions S_n and y_n by

$$S_n(x) = \begin{cases} \sup\{f(t): t\in[x_{i-1}, x_i]\} & \text{for } x = y_i, i = 1, 2, \ldots, 2n \\ \max\{S_n(y_i), S_n(y_{i+1})\} & \text{for } x = x_i, i = 1, 2, \ldots, 2n \\ S_n(0) = S_n(2\pi) \\ \text{linear and continuous for } x\in[x_{i-1}, y_i] \text{ and for} \\ x\in[y_i, x_i], i = 1, 2, \ldots, 2n \end{cases}$$

$$y_n(x) = \begin{cases} \inf\{f(t): t\in[x_{i-1}, x_i]\} & \text{for } x = y_i, i = 1, 2, \ldots, 2n \\ \min\{y_n(y_i), y_n(y_{i+1})\} & \text{for } x = x_i, i = 1, 2, \ldots, 2n \\ y_n(0) = y_n(2\pi) \\ \text{linear and continuous for } x\in[x_{i-1}, y_i] \text{ and for} \\ x\in[y_i, x_i], i = 1, 2, \ldots, 2n \end{cases}$$

Obviously

$$y_n(x) \leqslant f(x) \leqslant S_n(x) \quad x\in[0, 2\pi] \tag{8.11}$$

The derivatives S_n' and y_n' of the functions S_n and y_n exist in each point of the interval $[0, 2\pi]$ except eventually at the points $x_i, i = 0, 1, \ldots, 2n, y_i, i = 1, 2, \ldots, n$. Using the definitions of the functions S_n and y_n, we obtain

$$\begin{aligned} |S_n'(x)| &\leqslant 2n\pi^{-1}\omega(f, x; 4\pi n^{-1}) \quad x \neq x_i, y_i \\ |y_n'(x)| &\leqslant 2n\pi^{-1}\omega(f, x; 4\pi n^{-1}) \quad x \neq x_i, y_i \end{aligned} \tag{8.12}$$

If, for example, we let $x\in(y_i, x_i)$, then, since S_n is linear in (y_i, x_i), we have

$$|S_n'(x)| \leqslant 2n\pi^{-1}|S_n(y_{i+1}) - S_n(y_i)| \leqslant 2n\pi^{-1}\omega\left(f, x; \frac{4\pi}{n}\right)$$

On the other hand, again from the definition of the functions S_n and y_n, we obtain

$$0 \leqslant S_n(x) - y_n(x) \leqslant \omega\left(f, x; \frac{3\pi}{n}\right) \tag{8.13}$$

From (8.12) it follows that

$$\begin{aligned} \|S_n'\|_{L_p[0, 2\pi]} &\leqslant 2n\pi^{-1}\tau(f; 4\pi n^{-1})_{L_p} \\ \|y_n'\|_{L_p[0, 2\pi]} &\leqslant 2n\pi^{-1}\tau(f; 4\pi n^{-1})_{L_p} \end{aligned} \tag{8.14}$$

On the other hand, from (8.13) we get

$$\|S_n - y_n\|_{L_p} \leqslant \tau\left(f; \frac{3\pi}{n}\right)_{L_p} \tag{8.15}$$

From Theorem 8.1 it follows immediately (see Corollary 8.2) that for every absolutely continuous function g with $g'\in L_p$ we have

$$\tilde{E}_n(g)_{L_p} \leqslant \frac{2\pi}{n+1}\|g'\|_{L_p} \tag{8.16}$$

The inequalities (8.14) and (8.16) give us

$$\tilde{E}_n(S_n)_{L_p} \leqslant 4\tau\left(f;\frac{4\pi}{n}\right)_{L_p}$$

$$\tilde{E}_n(y_n)_{L_p} \leqslant 4\tau\left(f;\frac{4\pi}{n}\right)_{L_p} \tag{8.17}$$

The following inequality is obvious:

$$\tilde{E}_n(f)_{L_p} \leqslant \tilde{E}_n(S_n)_{L_p} + \|S_n - y_n\|_{L_p} + \tilde{E}_n(y_n)_{L_p} \tag{8.18}$$

Using property (5) of $\tau(f;\delta)_{L_p}$, from (8.15), (8.17) and (8.18) we obtain

$$E_n(f)_{L_p} \leqslant 4\tau\left(f;\frac{4\pi}{n}\right)_{L_p} + \tau\left(f;\frac{3\pi}{n}\right)_{L_p} + 4\tau\left(f;\frac{4\pi}{n}\right)_{L_p}$$

$$\leqslant 8(4\pi + 1)\tau(f;n^{-1})_{L_p} + (3\pi + 1)\tau(f;n^{-1})_{L_p}$$

$$\leqslant 120\tau(f;n^{-1})_{L_p} \qquad\blacksquare$$

Remark.　Of course, the constant 120 in Theorem 8.2 is not the best possible. The problem of obtaining the exact constant in Theorem 8.2 remains open.

At the end of this section we shall give an estimate for $\tilde{E}_n(f)_{L_p}$ using $\tau_k(f;n^{-1})_{L_p}$. First we shall need some lemmas.

Lemma 8.1.　*Let f be a bounded measurable 2π-periodic function. Let us consider $\omega_k(f,x;h)$ as a function of x with fixed k and h. Then*

$$\tau_1(\omega_k(f,x;h);\delta)_{L_p} \leqslant \tau_k\left(f;h+\frac{\delta}{k}\right)_{L_p}$$

Proof.　Let $g(x) = \omega_k(f,x;h)$. Since $g(x) \geqslant 0$ for all x,

$$\omega_1(g,x;\delta) = \sup\left\{|\Delta_\theta g(t)|: t, t+\theta \in \left[x - \frac{\delta}{2}, x + \frac{\delta}{2}\right]\right\}$$

$$= \sup\left\{|g(t+\theta) - g(t)|: t, t+\theta \in \left[x - \frac{\delta}{2}, x + \frac{\delta}{2}\right]\right\}$$

$$\leqslant \sup\left\{g(t): t \in \left[x - \frac{\delta}{2}, x + \frac{\delta}{2}\right]\right\}$$

$$= \sup\left\{\sup\left\{|\Delta_\mu^k f(s)|: s, s+k\mu \in \left[t - \frac{kh}{2}, t + \frac{kh}{2}\right]\right\}: t \in \left[x - \frac{\delta}{2}, x + \frac{\delta}{2}\right]\right\}$$

$$\leqslant \sup\left\{|\Delta_\mu^k f(t)|: s, s+k\mu \in \left[x - \frac{\delta}{2} - \frac{kh}{2}, x + \frac{\delta}{2} + \frac{kh}{2}\right]\right\} = \omega_k\left(f,x;h+\frac{\delta}{k}\right)$$

$\blacksquare$

Lemma 8.2.　*Let f be a bounded measurable 2π-periodic function and $g_n(x) =$*

$\omega_k(f, x; n^{-1})$. Then

$$\tilde{F}_n(g_n)_{L_p} \leqslant 120\tau_k(f; 2n^{-1})_{L_p}$$

Proof. From Lemma 8.1, letting $h = \delta = n^{-1}$, we have

$$\tau_1(g_n; n^{-1})_{L_p} \leqslant \tau_k(f; 2n^{-1})$$

From this inequality and Theorem 8.2 we obtain

$$\tilde{E}_n(g_n)_{L_p} \leqslant 120\tau_1(g_n; n^{-1})_{L_p} \leqslant 120\tau_k(f; 2n^{-1})_{L_p} \qquad \blacksquare$$

The following lemma is obvious.

Lemma 8.3. *Let f, g and φ be 2π-periodic bounded measurable functions and $f, g, \varphi \in L_p$. If for every x we have*

$$|f(x) - g(x)| \leqslant \varphi(x)$$

then

$$\tilde{E}_n(f)_{L_p} \leqslant \tilde{E}_n(g)_{L_p} + 2\tilde{E}_n(\varphi)_{L_p} + 2\|\varphi\|_{L_p}$$

Theorem 8.3. *Let f be a bounded measurable 2π-periodic function. For every natural number k there is a constant $c(k)$, dependent only on k, such that*

$$\tilde{E}_n(f)_{L_p} \leqslant c(k)\tau_k(f; n^{-1})_{L_p} \quad n > k$$

Proof. In view of Theorem 2.5, for every natural number k and every $h > 0$ there exists a function $f_{k,h} \in L_p[0, 2\pi]$ with the properties:

1. $|f(x) - f_{k,h}(x)| \leqslant \omega_k(f, x; 2h)$
2. $\|f - f_{k,h}\|_{L_p} \leqslant \omega_k(f; h)_{L_p}$
3. $\|f_{k,h}^{(k)}\|_{L_p} \leqslant c''(k)h^{-k}\omega_k(f; h)_{L_p}$

where $c''(k)$ is a constant, dependent only on k. From the proof of Theorem 2.5, in the periodic case we see that such a function can be, for example, the modified Steklov function.

Let $n > k$ and $h = n^{-1}$. Then, applying Lemma 8.3 to the functions f, $f_{k,h}$ and $\varphi(x) = \omega_k(f, x; 2n^{-1})$, in view of property (1) of the function $f_{k,h}$ and Lemma 8.2, we obtain

$$\tilde{E}_n(f)_{L_p} \leqslant \tilde{E}_n(f_{k,h})_{L_p} + 2\tilde{E}_n(\varphi)_{L_p} + 2\|\omega_k(f, \cdot; 2n^{-1})\|_{L_p}$$
$$\leqslant \tilde{E}_n(f_{k,h})_{L_p} + 240\tau_k(f; 4n^{-1})_{L_p} + 2\tau_k(f; 2n^{-1})_{L_p} \qquad (8.19)$$

To estimate $\tilde{E}_n(f_{k,h})_{L_p}$ we shall use Corollary 8.2 and property (3) of the function $f_{k,h}$. For $n > k$ we find that

$$\tilde{E}_n(f_{k,h})_{L_p} \leqslant c'(k)\|f_{k,h}^{(k)}\|_{L_p}n^{-k}$$
$$\leqslant c'(k)c''(k)\omega_k(f; n^{-1})_{L_p} \leqslant c'''(k)\tau_k(f; n^{-1})_{L_p} \qquad (8.20)$$

In view of property (5) of τ_k, from (8.19) and (8.20) we obtain the theorem.

$\blacksquare$

8.3.   Converse theorem for best one-sided approximations

We shall now obtain an analogue of the Salem–Stečkin estimate (8.2) for the averaged moduli of smoothness and best one-sided trigonometric approximations.

Theorem 8.4.   *Let f be a bounded measurable 2π-periodic function. For every natural number k there exists a constant $c(k)$ dependent only on k such that*

$$\tau_k(f;n^{-1})_{L_p} \leqslant \frac{c(k)}{n^k} \sum_{s=0}^{n} (s+1)^{k-1} \tilde{E}_s(f)_{L_p}$$

Proof.   For every natural number n let the trigonometric polynomials $P_n \in T_n$ and $Q_n \in T_n$ be such that

$$\tilde{E}_n(f)_{L_p} = \| P_n - Q_n \|_{L_p}$$
$$Q_n(x) \leqslant f(x) \leqslant P_n(x) \quad x \in [0, 2\pi]$$

Let $x \in [0, 2\pi]$ be fixed and $t, t + kh \in [x - k\delta/2, x + k\delta/2]$.
If $\Delta_h^k f(t) \geqslant 0$, then

$$0 \leqslant \Delta_h^k f(t) = \sum_{m=0}^{k} (-1)^m \binom{k}{m} f[t + (k-m)h]$$

$$\leqslant \sum_{i=0}^{[k/2]} \binom{k}{2i} P_n[t + (k-2i)h] - \sum_{i=0}^{[(k-1)/2]} \binom{k}{2i+1} Q_n[t + (k-2i-1)h]$$

$$= \Delta_h^k P_n(t) - \sum_{i=0}^{[(k-1)/2]} \binom{k}{2i+1} \{ Q_n[t + (k-2i-1)h]$$
$$- P_n[t + (k-2i-1)h] \}$$

$$= \Delta_h^k P_n(t) + \sum_{i=0}^{[(k-1)/2]} \binom{k}{2i+1} \{ P_n[t + (k-2i-1)h]$$
$$- Q_n[t + (k-2i-1)h] - [P_n(x) - Q_n(x)] \}$$

$$+ \sum_{i=0}^{[(k-1)/2]} \binom{k}{2i+1} [P_n(x) - Q_n(x)]$$

$$\leqslant \Delta_h^k P_n(t) + 2^k \{ \omega_1(P_n - Q_n, x; k\delta) + [P_n(x) - Q_n(x)] \}$$

i.e.

$$|\Delta_h^k f(t)| \leqslant |\Delta_h^k P_n(t)| + 2^k [\omega_1(P_n - Q_n, x; k\delta) + |P_n(x) - Q_n(x)|] \quad (8.21)$$

If $\Delta_h^k f(t) \leqslant 0$, then, in a similar way, we obtain

$$|\Delta_h^k f(t)| \leqslant |\Delta_h^k Q_n(t)| + 2^k [\omega_1(P_n - Q_n, x; k\delta) + |P_n(x) - Q_n(x)|] \quad (8.22)$$

From (8.21) and (8.22) we obtain

$$\omega_k(f, x; \delta) \leqslant \omega_k(P_n, x; \delta) + \omega_k(Q_n, x; \delta)$$
$$+ 2^k [\omega_1(P_n - Q_n, x; k\delta) + |P_n(x) - Q_n(x)|] \quad (8.23)$$

From this inequality it follows that

$$\tau_k(f;\delta)_{L_p} \leqslant \tau_k(P_n;\delta)_{L_p} + \tau_k(Q_n;\delta)_{L_p} + 2^k[\tau_1(P_n - Q_n;k\delta)_{L_p} + \tilde{E}_n(f)_{L_p}] \quad (8.24)$$

Since $\tau_1(\varphi;\delta)_{L_p} \leqslant \delta \| \varphi' \|_{L_p}$, using the Bernstein inequality $\| T' \|_{L_p} \leqslant n \| T \|_{L_p}$ for $T \in T_n$, we have

$$\tau_1(P_n - Q_n;k\delta)_{L_p} \leqslant k\delta \| P'_n - Q'_n \|_{L_p} \leqslant k\delta n \| P_n - Q_n \|_{L_p} = k\delta n \tilde{E}_n(f)_{L_p} \quad (8.25)$$

The inequalities (8.24) and (8.25) give us

$$\tau_k(f;\delta)_{L_p} \leqslant \tau_k(P_n;\delta)_{L_p} + \tau_k(Q_n;\delta)_{L_p} + 2^k(k\delta n + 1)\tilde{E}_n(f)_{L_p} \quad (8.26)$$

We now shall apply the standard method of Salem–Steckin. Let us set $n = 2^{s_0}$. Since

$$\tau_k(\varphi + \psi;\delta)_{L_p} \leqslant \tau_k(\varphi;\delta)_{L_p} + \tau_k(\psi;\delta)_{L_p}$$

then from (8.26) we obtain

$$\tau_k(f;\delta)_{L_p} \leqslant \sum_{i=1}^{s_0} [\tau_k(P_{2^i} - P_{2^{i-1}};\delta)_{L_p} + \tau_k(Q_{2^i} - Q_{2^{i-1}};\delta)_{L_p}]$$

$$+ \tau_k(P_1 - P_0;\delta)_{L_p} + \tau_k(Q_1 - Q_0;\delta)_{L_p} + 2^k(k\delta n + 1)\tilde{E}_n(f)_{L_p} \quad (8.27)$$

On the other hand, from property (6') of the averaged moduli of smoothness we obtain

$$\tau_k(P_{2^i} - P_{2^{i-1}};\delta)_{L_p} \leqslant k\delta^k \| (P_{2^i} - P_{2^{i-1}})^{(k)} \|_{L_p}$$
$$\leqslant k\delta^k 2^{ik} \| P_{2^i} - P_{2^{i-1}} \|_{L_p}$$
$$\leqslant k\delta^k 2^{ik} [\| P_{2^i} - f \|_{L_p} + \| P_{2^{i-1}} - f \|_{L_p}]$$
$$\leqslant k\delta^k 2^{ik} [\| P_{2^i} - Q_{2^i} \|_{L_p} + \| P_{2^{i-1}} - Q_{2^{i-1}} \|_{L_p}]$$
$$\leqslant 2k\delta^k 2^{ik} \tilde{E}_{2^{i-1}}(f)_{L_p} \quad (8.28)$$

and in the same way

$$\tau_k(Q_{2^i} - Q_{2^{i-1}};\delta)_{L_p} \leqslant 2k\delta^k 2^{ik} \tilde{E}_{2^{i-1}}(f)_{L_p} \quad (8.29)$$

From (8.27)–(8.29) we obtain

$$\tau_k(f;\delta)_{L_p} \leqslant 4k\delta^k \sum_{i=1}^{s_0} 2^{ik} \tilde{E}_{2^{i-1}}(f)_{L_p} + 2k\delta^k \tilde{E}_0(f)_{L_p} + 2^k(k\delta n + 1)\tilde{E}_n(f)_{L_p}$$

$$\leqslant 4^{k+1} k\delta^k \sum_{s=0}^{n} (s+1)^{k-1} \tilde{E}_s(f)_{L_p} + 2^k(k\delta n + 1)\tilde{E}_n(f)_{L_p} \quad (8.30)$$

Let $\delta = n^{-1}$. Then (8.30) gives us

$$\tau_k(f;\delta)_{L_p} \leqslant 4^{k+1} kn^{-k} \sum_{s=0}^{n} (s+1)^{k-1} \tilde{E}_s(f)_{L_p} + 2^k(k+1)\tilde{E}_n(f)_{L_p}$$

$$\leqslant 2^{3k+1} n^{-k} \sum_{s=0}^{n} (s+1)^{k-1} \tilde{E}_s(f)_{L_p}$$

which proves the theorem for $n = 2^{s_0}$. If $2^{s_0} \leqslant n < 2^{s_0+1}$, then

$$\tau_k(f;n^{-1})_{L_p} \leqslant \tau_k(f;2^{-s_0})_{L_p}$$

$$\leqslant 2^{3k+1}2^{-s_0 k}\sum_{s=0}^{2^{s_0}}(s+1)^{k-1}\tilde{E}_s(f)_{L_p}$$

$$\leqslant 2^{4k+1}n^{-k}\sum_{s=0}^{n}(s+1)^{k-1}\tilde{E}_s(f)_{L_p} \qquad \blacksquare$$

From Theorems 8.3 and 8.4 we obtain the following important statement.

Corollary 8.3. *For $0 < \alpha < k$ and $1 \leqslant p \leqslant \infty$ we have $\tau_k(f;\delta)_p = O(\delta^\alpha)$ if and only if $\tilde{E}_n(f)_p = O(n^{-\alpha})$.*

This corollary shows that best one-sided trigonometric approximations in L_p can be characterized by the averaged moduli of smoothness in the same way as the best L_p trigonometric approximations can be characterized by the usual moduli of smoothness in L_p.

Remark. Best one-sided approximations by splines can also be characterized by the averaged moduli of smoothness.

8.4. Notes

The constant 2π in Theorem 8.1 is exact. This was proved by Alexandrov [2].

One-sided approximation by algebraic polynomials is considered by Ivanov [2] and Stojanova [1].

One-sided approximation by entire functions of exponential type is considered by Drianov [1], [4], [5], where direct and converse theorems for the corresponding best one-sided approximation are obtained by means of the averaged moduli of smoothness (see also Taberski [2]).

Estimates for best one-sided approximations by means of the averaged moduli of smoothness of fractional type are obtained by Drianov [3].

Direct and converse theorems for best one-sided approximations in L_p, $0 < p < 1$ using the corresponding averaged moduli of smoothness are obtained by Šardin [1] (see also Drianov [7] and Taberski [3]).

Characterization of the spaces A_{pq}^α (see the notes to Chapter 2) is obtained by Popov [10]:

$$f \in A_{pq}^\alpha \quad \text{iff} \quad \left\{\sum_{n=0}^{\infty}[2^{n\alpha}\tilde{E}_{2^n}(f)_{L_p}]^q\right\}^{1/q} < \infty$$

One-sided algebraic approximation on the real line with the weight $\exp(-x^2)$ is considered by Ivanov [3].

Petrushev [1] used the first averaged modulus for obtaining an estimate for best Hausdorff approximations by rational functions.

Dolzenko [1] used the first averaged modulus for Hausdorff approximations by piecewise monotone functions.

Multivariate one-sided approximations are considered in Popov [8], [11] and Hristov, Ivanov [1].

REFERENCES

Ahlberg, J. H., Nilson, E. N. and Walsh J. L.

1. *The Theory of Splines and their Applications*, Academic Press, New York and London, 1967.

Alexandrov, L. G.

1. Error estimations for the numerical solution of elliptic equations. *Constructive Theory of Functions' 84*, Publ. House of Bulg. Acad. of Sci., Sofia, 1984, 106–112.

2. The best possible constant in the relation between the best one-sided approximation of an absolutely continuous function and the best approximation of its derivative. *C.R. Acad. bulg. des Sci.*, **38**, 2, 1985, 191–193.

3. On the traces of the functions which belong to the spaces generated by the averaged moduli of smoothness. *C.R. Acad. bulg. des Sci.*, **38**, 11, 1985, 1457–1460.

Andreev, A. S.

1. Some estimations for the approximate solution of differential equations. *Approximation and Function Spaces*, Z. Ciesielski (ed.), PWN–Warszawa–North-Holland Publ. Company, Amsterdam–New-York–Oxford, 1981, 16–24.

2. Convergence rate for spline collocation to the Fredholm integral equation of the second kind. *Pliska, Studia math. bulg.*, **5**, 1983, 84–92.

3. Interpolation by quadratic and cubic splines in L_p. *Constructive Function Theory '81*, Publ. House of Bulg. Acad. of Sci., Sofia, 1983, 211–216.

4. Error estimates for the numerical solution of the Cauchy problem. *Serdica, Bulg. math. publ.*, **10**, 1984, 206–215.

Andreev, A. S. and Gichev, D.

1. Estimates for the error of finite-difference methods for the solution of the Cauchy problem. *Numerical Methods and Applications '84*, Publ. House of Bulg. Acad. of Sci., Sofia, 1985, 152–156.

Andreev, A. S. and Popov, V. A.

1. Approximation of functions by means of linear summation operators in L_p. *Colloquia Mathematica Societatis János Bolyaj*, **35**, 1980, 127–150.

Andreev, A. S., Popov, V. A. and Sendov, Bl.

1. Jackson's type theorems for one-sided polynomial and spline approximation. *C.R. Acad. bulg. des Sci.*, **30**, 11, 1977, 1533–1536.

2. Jackson type theorems for the best one-sided approximation by means of trigonometric polynomials and splines. *Math. Notes*, **26**, 5, 1979, 791–804 (in Russian).

3. Estimates for the error of the numerical solution of ordinary differential equations, *Journal Vichisl. Math. and Math. Fisics*, **21**, 3, 1981, 635–650 (in Russian).

Babenko, V. F. and Ligun, A. A.

1. The order of the best one-sided approximations by polynomials and splines in the L_p-metric. *Math. Notes*, **19**, 1976, 323–329 (in Russian).

174 References

Bahvalov, N. S.
1. *Numerical Methods*, vol. I, Moscow, 1973 (in Russian).
Berezin, I. S. and Zhidkov, N. P.
1. *Methods of Calculations*, vol. I, Moscow 1962, vol. II, 1960 (in Russian).
Bergh, J. and Löfström, J.
1. *Interpolation Spaces. An Introduction.* Springer-Verlag, Berlin–Heidelberg, 1976.
Binev, P. G.
1. Superconvergence in spline interpolation. *C.R. Acad, bulg. des Sci.*, **37**, 12, 1984, 1613–1616.
2. Superconvergence in the spline-interpolation. *Constructive Theory of Functions '84*, Publ. House of Bulg. Acad. of Sci., Sofia, 1984, 164–170.
3. Spline interpolation of bounded functions in the class $L_p[0, 1]$. *Serdica, Bulg. math. publ.*, **11**, 1985, 42–47.
4. Convergence and superconvergence in Hermite spline interpolation. *Numerical Methods and Applications '84*, Publ. House of Bulg. Acad. of Sci., Sofia, 1985, 179–184.
5. $O(n)$ bounds of Whitney constants. *C.R. Acad. bulg. des. Sci.*, **38**, 1985, 1303–1305.
Brudnii, J. A.
1. Approximation of functions of n-variable quasipolynomials. *Isv. Acad. Nauk SSSR, seria matem.*, **34**, 1970, 564–583 (in Russian).
Butzer, P. L., Dyckoff, H., Görlich, E. and Stens, R. L.,
1. *Canad. J. Math.*, **29**, 1977, 781–793.
Djukanova, T. A.
1. Approximation of functions by means of linear summation Baskakov's operators in L_p. *Pliska, Studia Math. bulg.* **5**, 1983, 32–39.
Dolzenko, E. P.
1. Some new inverse theorems in approximation theory. *Constructive theory of Functions '77*, Publ. House of Bulg. Acad. of Sci., Sofia, 1980, 49–56 (in Russian).
Dolzenko, E. P. and Sevastjanov, E. A.
1. On the approximation of functions in a Hausdorff metric by means of piecewise monotone (in particular rational) functions. *Mat. Sbornik*, **101**, 1976, 503–541 (in Russian).
Doronin, V. G. and Ligun, A. A.
1. Upper bounds for the best one-sided approximations of the classes $W^r L_1$ by splines. *Math. Notes*, **19**, 1976, 11–17 (in Russian).
Drianov, D. P.
1. Converse theorem of one-sided approximation with entire functions of exponential type. *C.R. Acad. bulg. des Sci.*, **34**, 9, 1981, 1233–1236.
2. Averaged modulus of smoothness of fractional index and fractional order derivatives. *C.R. Acad. bulg. des Sci.*, **35**, 12, 1982, 1635–1637.
3. Averaged modulus of smoothness of fractional index and applications. *C.R. Acad. bulg. des Sci.*, **36**, 1, 1983, 41–43.
4 One-sided approximation with entire functions of exponential type. *Serdica, Bulg. math. publ.*, **10**, 1984, 276–286.
5. Direct and converse theorems for one-sided approximation by entire functions. *C.R. Acad. bulg. des Sci.*, **39**, 2, 1986, 23–25.
6. On the convergence and saturation problem of a sequence of discrete linear operators of exponential type in $L_p(-\infty, \infty)$ spaces. *Acta Math. Hung.*, **49** (1–2), 1987, 103–127.
7. One-sided approximation by trigonometric polynomials in L_p-metric, $0 < p < 1$. *Proc. of Banach Centre*, 1986 (to appear).
Dzjadik, V. K.
1. *Introduction to the theory of uniform approximation of functions by polynomials*, Moscow, 1977.
Fedotova, O. A.
1. Variational difference scheme for one-dimensional diffusian equation. *Math. Notes*, **17**, 1975, 893–898 (in Russian).

Freud, G.

1. Über einseitige Approximation durch Polynome. *I. Acta Sci. Math.*, **16**, 1955, 12–18.

Freud, G. and Popov, V. A.

1. Some questions connected with spline and polynomial approximation. *Studia Sci. Math. Hung.*, **5**, 1970, 161–171 (in Russian).

Ganelius, T.

1. On one-sided approximation by trigonometrical polynomials. *Math. Scandinavica*, **4**, 1956, 247–258.

Hoeffding, W.

1. The L_1-norm of the approximation error for Bernstein-type polynomials. *J. Approx. Theory*, **4**, 1971, 347–356.

Hristov, V. H.

1. A note for the convergence of one quadrature process. *C.R. Acad. bulg. des Sci.*, **36**, 9, 1983, 1151–1154 (in Russian).

2. On the coefficients of Fourier–Lagrange. *Pliska, Studia math. bulg.*, **5**, 1983, 23–31 (in Russian).

3. On the convergence in the mean of interpolation polynomials of periodic functions. *Pliska, Studia math. bulg.*, **5**, 1983, 14–22 (in Russian).

4. On the convergence of some interpolation processes in integral and discrete norms. *Constructive Function Theory '81*, Publ. House of bulg. Acad. of Sci., Sofia, 1983, 185–188 (in Russian).

5. On the quantitative Korovkin theorem in L_p for periodic functions. *C.R. Acad. bulg. des Sci.*, **36**, 8, 1983, 1035–1038 (in Russian).

6. Spaces of functions, generated by the averaged moduli of functions of many variables. *Constructive Theory of Functions '84*, Publ. House of Bulg., Acad. of Sci., Sofia, 1984, 97–101 (in Russian).

7. Connection between the usual and the averaged moduli of smoothness for multivariate functions. *C.R. Acad. bulg. des Sci.*, **38**, 2, 1985, 175–178 (in Russian).

8. Connection between Besov spaces and the spaces generated by the averaged moduli of smoothness in R^n. *C.R. Acad. bulg. des Sci.*, **38**, 5, 1985, 555–558 (in Russian).

Hristov, V. H., Ivanov, K. G.

1. Operators for one sided approximation of functions. *Constructive theory of functions '87*, Publ. House of Bulg. Acad. of Sci., Sofia, 1988, 222–232.

Ivanov, K. G.

1. New estimates of errors of quadrature formulae, formulae of numerical differentiation and interpolation. *Analysis Mathematica*, **6**, 1980, 281–303.

2. On the one-sided algebraic approximation in L_p. *C.R. Acad. bulg. des Sci.*, **32**, 8, 1979, 1037–1040.

3. One-sided algebraic approximation on the real axis with the weight $\exp(-x^2)$. *Serdica, Bulg. math. publ.*, **7**, 1981, 57–65.

4. On the rates of convergence of two moduli of functions. *Pliska, Studia math. bulg.*, **5**, 1983, 97–104.

5. Approximation by Bernstein polynomials in an L_p metric. *Constructive Theory of Functions '84*, Publ. House of Bulg. Acad. of Sci., Sofia, 1984, 421–429.

6. On the behaviour of two moduli of functions. *C.R. Acad. bulg. des Sci.*, **38**, 5, 1985, 539–542.

7. On the behaviour of two moduli of functions. II. *Serdica, Bulg. Math. publ.*, **12**, 1986, 196–203.

8. Converse theorems for approximation by Bernstein polynomials in $L_p[0,1]$ ($1 < p < \infty$). *Constructive Approximation*, **2**, 1986, 377–392.

Ivanov, K. G. and Takev, M.

1. $O(n \ln n)$ bounds of Whitney constants. *C.R. Acad. bulg. des Sci.*, **38**, 1985, 1129–1135.

Korovkin, P.P.

1. *Linear Operators in the Theory of Approximation*, Moscow, 1959 (in Russian).

2. An experience of axiomatic construction of some problems of the theory of

176 **References**

approximation. *Uch. Zapiski Kalinin. Ped. Inst.*, **69**, 1969, 91–109 (in Russian).
Lorentz, G. G.
1. Approximation of Functions, Holt, New York, 1966.
Mamedov, R. G.
1. On the order of approximation of functions by linear positive operators. *Dokladi AN SSSR*, **128**, 11, 1959, 674–676 (in Russian).
Mevissen, H., Nessel, R. J. and Wickeren, E. van.
1. On the Riemann convergence of positive linear operators. *Preprint RWTH Aachen*, **332**, 1986.
Musielak, H.
1. On the τ-modulus of smoothness in some functional spaces. *Constructive Function Theory '81* Publ. House of Bulg. Acad. of Sci., Sofia, 1983, 447–454.
Natanson, I. P.
1. *Constructive function theory*. Moscow, 1949 (in Russian).
Niederreiter, H.
1. Methods for estimating descrepancy. *Proc. Sympos. of Appl. on Number to Numeric Analysis, Montreal 1971*, New York, 1972.
Nikolskii, S. M.
1. *Quadrature formulae*. Moscow, 1974.
Petrushev, P. P.
1. Best rational approximations in a Hausdorff metric. *Serdica Bulg. Math. publ.*, **6**, 1980, 29–41 (in Russian).
Popov, V. A.
1. Remark on the one-sided approximation of functions. *C.R. Acad. bulg. des Sci.*, **32**, 10, 1979, 1319–1322 (in Russian).
2. Some problems of estimation of the error of numerical solution of differential equations. *Variational Difference Methods in Mathematical Physics*, Novosibirsk, 1981, 105–112 (in Rusian).
3. Converse theorem for one-sided trigonometric approximation. *C.R. Acad. bulg. des Sci.*, **30**, 1977, 1529–1532.
4. Direct and Converse Theorems for One-sided Approximation, ISNM 40, Birkhäuser-Verlag Basel, 1978, 449–458.
5. On the one-sided approximation of functions. *Constructive Function Theory '77*, Publ. House of Bulg. Acad. of Sci., Sofia, 1980, 465–468.
6. Averaged local moduli and their applications. *Approximation and Function Spaces*, Z. Ciesielski (ed.), PWN–Warszawa–North-Holland Publ. Company, Amsterdam–New York–Oxford, 1981, 572-583.
7. On the quantitative Korovkin theorems in L_p. *C.R. Acad. bulg. des Sci.*, **35**, 7, 1982, 897–900.
8. One-sided approximation of periodic functions of several variables. *C.R. Acad. bulg. des Sci.*, **35**, 12, 1982, 1639–1642.
9. Averaged moduli and their function spaces. *Constructive Function Theory '81*, Publ. House of Bulg. Acad. of Sci., Sofia, 1983, 482–487.
10. Function spaces, generated by the averaged moduli of smoothness. *Pliska, Studia math. bulg.*, **5**, 1983, 132–143.
11. On the one-sided approximation of multivariate functions. *Approximation Theory IV*, Academic Press, 1983, 657–661.
12. The one-sided K-functional and its interpolation spaces. *Proc. of the Steklov Inst of Math.*, **4**, 1985, 229–232.
Popov, V. A. and Andreev, A. S.
1. Steckin type theorems for one-sided trigonometric and spline approximation. *C.R. Acad. bulg. des. Sci.*, **31**, 4, 1978, 151–154.
2. On the error estimation in numerical methods. *Computational Mathematics*, vol.

13, Banach Centre Publications, PWN–Warsaw, 1984, 647–658.

Popov, V. A. and Dechevski, L. T.

1. On the error of numerical solution of parabolic equations in network norms. *C.R. Acad. bulg. des Sci.*, **36**, 4, 1984, 429–432.

Popov, V. A. and Hristov, V. H.

1. Averaged moduli of smoothness for functions of several variables, and the function spaces generated by them. *Proc. of Steklov Inst. of Math.*, Issue 2, 1985, 155–160.

Popov, V. A. and Szabados, J.

1. On the convergence and saturation of the Jackson polynomials. *J. Approx. Theory Appl.*, 1, 1984, 1–10.

Popov, P.

1. *Constructive Function Theory '77*, Publ. House of Bulg. Acad of Sci., Sofia, 1980, 121–12.

Proinov, P. D.

1. Upper estimation of the error of the general quadrature process with positive weights. *C.R. Acad. bulg. des Sci.*, **35**, 1982, 605–608 (in Russian).

Quak, E.

1. On the connection of multivariate τ-moduli and Bernstein polynomials. *C.R. Acad. bulg. des Sci.*, **40**, 5, 1987, 17–19.

Samarskii, A. A.

1. *Introduction to the Theory of Difference Schemes.* Moscow, 1974 (in Russian).

2. *Theory of Difference Schemes.* Moskow, 1977 (in Russian).

Sardin, A.

1. The order of one-sided approximation of functions in an L_p metric. *Analysis Math.*, **12**, 1986, 175–184.

Sendov, Bl.

1. Approximation with respect to the Hausdorff distance. Dissertation, Moscow, 1967 (in Russian).

2. Convergence of sequences of monotonic operators in A-distance. *C.R. Acad. bulg. des Sci.*, **30**, 1977, 657–660.

3. *Hausdorff Approximations.* Publ. House of Bulg. Acad. of Sci., Sofia, 1979 (in Russian).

4. On the constants of H. Whitney. *C.R. Acad. bulg. des Sci.*, **35**, 1982, 431–434.

5. The constants of H. Whitney are bounded. *C.R. Acad. bulg. des Sci.*, **38**, 1985, 1209–1302.

6. On the theorem and constants of H. Whitney. *Constructive Approximation*, **3**, 1987, 1–11.

7. Modification of the Steklov function. *C.R. Acad. bulg. des Sci.*, **36**, 1983, 315–317.

Sendov, Bl. and Popov, V. A.

1. *Numerical Methods*, vol. I, Narodna prosveta, Sofia, 1975, vol. II 1978 (in Bulgarian).

Stećkin, S. B.

1. On the order of the best approximations of continuous functions. *Isv. Acad. Nauk SSSR, seria metam.*, **15**, 1951, 219–242 (in Russian).

Stećkin, S B. and Subbotin, J.

1. *Splines in Computational Mathematics*, Moscow, 1976 (in Russian).

Stein, E. M. and Weiss, G.

1. *Introduction to Fourier Analysis on Euclidean Spaces*, Princeton University Press, Princeton, New Jersey, 1971.

Stojanova, M.

1. Characterization of the best one-sided approximation by algebraic polynomials in L_p. *Mathematica Balcanica*, **1**, 1987. (to appear).

Taberski, R.

1. On the modified integral moduli of smoothness and one-sided approximation of periodic functions. *Functiones et Approximatio*, **10**, 1980, 147–155.

2. One-sided approximation by entire functions. *Demonstratio Math.*, **15**, 2, 1982, 477–505.

3. One-sided trigonometric approximation in metrics of the Fréchet spaces $L_p(0 < p < 1)$. *Math. Nachr.*, **123**, 1985, 39–46.

Timan A. F.

1. *Theory of Approximation of Functions of a Real Variable*, Moscow, 1960.

Totkov, G.

1. On the approximation of bounded periodic functions in an L_p-metric, $1 \leqslant p \leqslant \infty$. *C.R Acad. bulg. des Sci.*, **35**, 10, 1982, 1357–1360 (in Russian).

2. Averaged moduli of smoothness, variation and integral inequalities in $R^s, s \geqslant 2$. *C.R. Acad. bulg. des Sci.*, **16**, 4, 1983, 452–456 (in Russian).

3. On the convergence of multivariate quadrature formulae. *C.R. Acad. bulg. des. Sci.*, **37**, 8, 1984, 1171–1174 (in Russian).

Totkov, G. and Baselkov, M.

1. On the convergence of some quadrature processes. *Constructive Theory of Functions '84*, Publ. House of Bulg. Acad. of Sci., Sofia, 1984, 864–869.

Whitney, H.

1. On functions with bounded nth difference. *J. Math. Pure & Appl.*, **36**, 1957, 67–95.

2. On bounded functions with bounded nth difference. *Proc. Amer. Math. Soc.*, **10**, 1959, 480–481.

Wickeren E. van

1. On the approximation of Riemann integrable functions by Bernstein polynomials. *Preprint RWTH Aachen*, **339**, 1987.

Zhidkov, E. P. Andreev, A. S. and Popov, V. A.

1. Gibbs effects for spline interpolation and for solving integral equations by the spline-collocation method. *Serdica, Bulg, math. publ.*, **12**, 1986, 315–320 (in Russian).

Zygmund, A.

1. *Trigonometrical Series*, 2nd ed., Cambridge University Press, Cambridge, 1959.

INDEX

179